Cryopreservation - Current Advances and Evaluations

Edited by Marian Quain

Published in London, United Kingdom

IntechOpen

Supporting open minds since 2005

Cryopreservation - Current Advances and Evaluations
http://dx.doi.org/10.5772/intechopen.77775
Edited by Marian Quain

Contributors
Jonathan Lakey, Samuel Rodriguez, David Whaley, Michael Alexander, Reza Mohammadi, Nabil Sayme, Dayong Gao, Ali Erdem Öztürk, Mustafa Numan Bucak, Mustafa Bodu, Nuri Başpınar, İlhami Çelik, Zhiquan Shu, Nazan Keskin, Mark D. Scott, Elisabeth Maurer-Spurej, Nobu Nakane, Samuel Buff, Pierre Bruyere, Anne Baudot, Hugo Desnos, Magda Teixeira, Gérard Louis, Stéphane Trombotto, Laurent David, Amani Moussa, Loris Commin, Miloš Faltus, Stacy Hammond, Jiří Zámečník, Noha Al-Otaibi, Shao Yuting, Chen Chao, Bi Yanglong, Qi Zhou, Jun Yang, Xiao Lv, Mingyue Lin

First published in London, United Kingdom, 2020 by IntechOpen
IntechOpen is the global imprint of INTECHOPEN LIMITED, registered in England and Wales, registration number: 11086078, 7th floor, 10 Lower Thames Street, London, EC3R 6AF, United Kingdom
Printed in Croatia

British Library Cataloguing-in-Publication Data
A catalogue record for this book is available from the British Library

Additional hard and PDF copies can be obtained from orders@intechopen.com

Cryopreservation - Current Advances and Evaluations
Edited by Marian Quain
p. cm.
Print ISBN 978-1-83880-205-9
Online ISBN 978-1-83880-206-6
eBook (PDF) ISBN 978-1-83880-630-9

For EU product safety concerns:
IN TECH d.o.o., Prolaz Marije Krucifikse Kozulić 3, 51000 Rijeka, Croatia, info@intechopen.com or visit our website at intechopen.com.

Meet the editor

Marian D. Quain is a biotechnologist whose research focuses on utilization of tissue culture techniques for production of clean planting materials, germplasm conservation and use of molecular tools for crop diversity, disease diagnostics and molecular marker assisted selection breeding using transcriptomics and proteomics, as well as application of recombinant gene technology for crop improvement. Marian was a Visiting Research Fellow at Leeds University, UK, from 2011 to 2013. In 2013, she was adjudged the National Best Research Scientist by the Ghanaian Ministry of Food and Agriculture (MoFA). She is a 2017 Cochran Fellow and a fellow of the Africa Science Leadership Program. In 2018, she won the American Society for Plant Biologists Award for Excellence in Education for her outstanding contributions to plant biology education.

Contents

Preface XI

Section 1
Recceing Oligomers, Polymers and Other Cryoprotective Agents 1

Chapter 1 3
The Use of Chitooligosaccharides in Cryopreservation: Discussion of Concept and First Answers from DSC Thermal Analysis
by Hugo Desnos, Pierre Bruyère, Magda Teixeira, Loris Commin, Gérard Louis, Stephane Trombotto, Amani Moussa, Laurent David, Samuel Buff and Anne Baudot

Chapter 2 29
Cryoprotection of Platelets by Grafted Polymers
by Mark D. Scott, Nobu Nakane and Elisabeth Maurer-Spurej

Chapter 3 49
Cryomedia Formula: Cellular Molecular Perspective
by Noha A. Al-Otaibi

Section 2
Procedures for Cryopreserving Gametes 61

Chapter 4 63
Cryopreservation of Human Spermatozoa: A New Frontier in Reproductive Medicine
by Nabil Sayme

Chapter 5 75
Cryobiology and Cryopreservation of Sperm
by Ali Erdem Öztürk, Mustafa Numan Bucak, Mustafa Bodu, Nuri Başpınar, İlhami Çelik, Zhiquan Shu, Nazan Keskin and Dayong Gao

Section 3
Cryobiology Aiding Organ Transplant 117

Chapter 6 119
Current Advancements in Pancreatic Islet Cryopreservation Techniques
by Samuel Rodriguez, David Whaley, Michael Alexander, Mohammad Rezaa Mohammadi and Jonathan R.T. Lakey

Chapter 7 **139**
Cryopreservation in Ophthalmology
by Yuting Shao, Chao Chen, Qi Zhou, Jun Yang, Xiao Lv, Mingyue Lin and Yanlong Bi

Section 4
Dynamics of Water Content in Plant Tissues During Cooling and Heating **159**

Chapter 8 **161**
Methods of Thermal Analysis as a Tool to Develop Cryopreservation Protocols of Vegetatively Propagated Crops
by Stacy D. Hammond, Miloš Faltus and Jiří Zámečník

Preface

Cryopreservation is the storage of biological material at ultra-low temperatures, preferably that of liquid nitrogen, which arrests all metabolic activities. This technique is widely applied to all organisms ranging from microorganisms through plants to animals and human organs. Theoretically, no genetic changes should occur during cryopreservation, thus permitting indefinite cell preservation. **Cryopreservation** or **cryoconservation** is a process where organelles, cells, tissues, extracellular matrices, organs, or any other biological constructs susceptible to damage caused by unregulated chemical kinetics are preserved by cooling to very low temperatures ranging from -80 to -210°C using programmable freezers, liquid nitrogen, and liquid nitrogen slush and vapour. A successful cryopreservation procedure therefore entails ensuring the normal functionality of the organism. In recent times, basic cryopreservation has had applications in research and clinical, medical, and agricultural fields.

Typically, freezing or cooling is lethal to most living organisms' cellular, mechanical, and metabolic functionality. Hence the typical procedures involved in cryopreservation include selecting appropriate tissue, conditioning the tissue, determining cooling rate, storing the tissue, thawing the tissue, and subsequently restoring tissue functions. During cryopreservation, several dynamics come into play; these include cell type, size, water content, temperature, and metabolic state. There have been several limitations to the effective application of cryopreservation techniques and storage, although substantial successes have been reported for both animal and plant cells. Cells have to be protected from damage, especially by ice crystals during freezing. Cryoprotectants, also called Cryo Protective Agents (CPAs), are widely used; however, these may be toxic to the cells depending on the type of cell and the extent of exposure. Mode of freezing, thawing, regenerating, and rejuvenating also may render the process successful or otherwise. Viability of cells may be compromised at any step of the process. Naturally, tissues subjected to cryotemperatures contain fluids such as intracellular and extracellular fluids, plasma, interstitial fluid, and transcellular fluid, which make up 10 to 75 percent of tissues. During cooling, these fluids may form ice crystals, which may nucleate to larger crystals and ultimately damage cells. Hence the use of CPAs may render stability and protection to cell membranes, and where proven useful, preserve functionality of tissue as subjected to cryopreservation.

The main objective of this book is to bring to bear factors that affect cells during cryopreservation. Divided into four sections, this book contains eight chapters providing discussions, overviews, and reviews of cryo-techniques.

The first section is on "Recceing Oligomers, Polymers, and Other Cryoprotective Agents (CPAs)." Successful cooling is dependent on the control of extracellular ice formation, protective intracellular dehydration, and the colligative and dehydration properties of the CPA. Hence CPAs ensure a freeze-avoidance mechanism that enables hydrated tissues to survive when exposed to cryogenic temperatures. There are, however, unanswered questions about the precise mechanism of action of CPAs. In their function to protect living cells against damage during cooling, they act by lowering the freezing point, modifying the crystalline surface of ice, and averting solute accumulation at a given level of dehydration, all of which counteract injury. The use of CPAs is an essential aspect in all cryopreservation protocols, from slow freezing to vitrification. There are several types including non-permeable (or external) and permeable (internal) CPAs, depending on their ability to permeate biological membranes. Some examples of non-permeable CPAs are glucose, sucrose, or polyvinylpyrrolidone (PVP). Dimethyl sulfoxide (DMSO), 1,2-propanediol, glycerol, and ethylene glycol are examples of permeable CPAs.

In the first chapter, "The Use of Chitooligosaccharides in Cryopreservation: Discussion of Concept and First Answers from DSC Thermal Analysis" Dr. Buff et al. highlight that a routinely used CPA is cytotoxic and apparently mutagenic as utilized in slow-freezing solutions. The authors thus propose new non-penetrating CPAs: chitooligosaccharides (COS). These chitosan oligomers molecules are biocompatible, antioxidant, and bacteriostatic. In their chapter, the authors test the probability of reducing penetrating CPAs during slow-freezing procedures. COS are thus proposed as extracellular CPAs that reduce the use of dimethyl sulfoxide (Me2SO). The authors question the biocompatibility of COS on mouse embryos through the analysis of the cells' development. They evaluate the molecules in slow-freezing solutions with a reduced quantity of Me2SO, and use differential scanning calorimetry (DSC) to evaluate the crystallization and melting processes, the amount of crystallized water, the equilibrium temperature, and consequently the impact of different CPAs. In this study, the authors indicated that COS could promote a decrease in the use of penetrating CPAs while ensuring successful vitrification. However, there remains the need for further investigation of the influence of COS on the organization of ice crystals in cryopreserved samples and the ability of COS to increase the glass-forming tendency of the remaining solution.

Further to this, Dr. Mark Scott et al. provide a chapter on "Cryoprotection of Platelets by Grafted Polymers." Platelets are lifesaving cells that are difficult to store at 4°C due to the formation of cold storage lesions (CSLs). The deployment of methoxypolyethylene glycol (mPEG), a biocompatible polymer, to the membrane of platelets reduces CSLs. In their study, the authors demonstrate that the covalently grafting platelets to mPEG, termed as PEGylation, serves as a potent CPA allowing platelets to be stored at 4°C, or frozen at −20°C, while maintaining normal hemostatic function and counts. Also, there is the prevention of the formation of overt morphological changes resulting from the CSLs. Ultimately the successful cold storage of platelets improves transfusion safety as it reduces the threat of microbial growth in a blood product to be used for patients. As already discussed, the prolonged used of DMSO is toxic to cells. The exciting news is that the use of frozen platelets does not require DMSO inclusion. Hence this requires that studies are conducted for protocol validation and to clinically ensure that PEGylation can facilitate the cold storage of platelets.

The following chapter by Noha A. Al-Otaibi considers "Cryomedia Formula: Cellular Molecular Perspective." The importance of CPAs cannot be over emphasised since the growing market of cell therapy medicinal products (CTMPs) and biopharmaceuticals demands efficient cryopreservation and CPAs. The review presented here considers topics such as conventional cryomedia compositions and protection mechanisms, quality assessment methods of cryopreserved cells, CPAs' protection action, CPA toxicity and detrimental effects, other biochemical effects, and modulating CPAs' damages via additive agents. In the conclusion, the author states, that "Understanding of the protective mechanisms of cryomedia ingredients along with identifying powerful protective compounds to enhance cryomedia performance is highly demanded."

The next section of the book, "Procedures for Cryopreserving Gametes," sheds light on the cryopreservation of sperm. In 1954, there was a report of using cryopreservation in humans where previously frozen sperms were inseminated resulting in pregnancies. Following this, in 1957, fowl sperms were cryopreserved by a team of scientists in the United Kingdom directed by Christopher Polge. One may be wondering about the need to use cryopreservation in human reproduction; will we be playing God? The answer is no, because this is simply a means of assisting natural processes. There may be times that it is clinically safe for the reproductive cell of an individual to be preserved for posterity, especially when undergoing, for example, chemotherapy, which may be detrimental to ones' reproductive abilities in the future. This is just one example, but there are several others. In the first chapter in this section, "Cryobiology and Cryopreservation of Sperm," Ali Erdem Öztürk et al. discuss the hypothesis of cryoinjury during cryopreservation and the associated challenges, especially damage caused by CPAs. As usual, DMSO plays a key role here where damage to DNA and mitochondria as well as formation of reducing oxygen species (ROS) are inevitable. This chapter raises issues that should be further investigated. The chapter is also closely related to the preceding chapter in which the authors discuss the use of less damaging CPAs including oligomers and polymers. The current chapter comprehensively addresses the biology, necrosis, and apoptosis of sperm cells during freezing with data supported by electron microscope images. The discussion provided is a guide to future research and helps to identify critical gaps that must be filled to ensure the full potential of using cryopreservation in enriching species' reproducibility.

In the next chapter, "Cryopreservation of Human Spermatozoa: A New Frontier in Reproductive Medicine," the authors present the imperative function of cryopreservation in assisted reproduction techniques. The need for improved techniques and functionality of sperms, including motility after thawing, is still a challenge although this system has been around for several years. Though new methods including lyophilization have been proposed, they need to be validated, and hence future research needs to investigate and optimize safety methods as well as health status of offspring.

There is the obvious clinical demand that during surgery where organs have to be isolated and transplanted, the isolated organ should be safely transported, stored, and be functional for a successful use. Technically, where distance and time are not a challenge, this can be effectively carried out, otherwise organs go to waste. It is necessary that demand and supply systems are established with the aim of stabilizing the biological tissues and preventing metabolic and biochemical processes that render organs nonfunctional. The next section of the book, "Cryobiology Aiding to Organ Transplant," focuses on procedures that have to be improved so that sub-zero temperatures can assist in saving lives. In the first chapter of this section, "Cryopreservation in Ophthalmology," Dr. Shao et al. review how cryopreservation can be deployed to facilitate amniotic membrane and cornea preservation in ophthalmology. There is more emphasis on CPAs in this chapter where glycerol-cryopreserved cornea tissues can be effectively and bio-safely used. The need for equipment, however, complicates the high cost of technical support required. Transportation is still a bottleneck that has to be dealt with to maintain functioning properties of corneal endothelial cells.

In the next chapter, "Current Advancements in Pancreatic Islet Cryopreservation Techniques," Dr. Rodriguez et al. outline the limitations of effective storage during donor-recipient cross-matching after islet isolation. The authors characterize aspects of the islet cryopreservation method under the following topics: history of islet cryopreservation and characteristics of the islet cryopreservation process such as cryoprotection and thawing. They present advances in islet CPA technology with a focus on permeating and non-permeating reagents, as well as the use of alginate-based microencapsulation technology. In conclusion, the authors are hopeful that a standard cryopreservation protocol, islet banking, would be more feasible and that eventually transportation "would no longer be throttled by the donor-recipient mismatch."

The final section of the book is titled "Dynamics of Water Content in Plant Tissues during Cooling and Heating." Water is a critical component of organisms and serves as the medium for metabolic activities occurring inter and intracellularly. Water thus gets in the way when cells need to be cooled, prompting the questions, how much water should a cell lose and how should this be done to maintain viability and genetic integrity of the organism? In the chapter, "Methods of Thermal Analysis as a Tool to Develop Cryopreservation Protocols of Vegetatively Propagated Crops," Dr. Hammond et al. place emphasis on the challenge of reducing water content of vegetative plant tissues to facilitate successful cryotreatment. It is critical that glass transition is formed during cooling because if ice crystals are formed, the cells are damaged. The application of DSC, temperature modulation DSC, and quasi-isothermal modulation DSC (QITTMDSC) are used here to monitor heat flow types in measuring the quantity of freezable water. The authors conclude by highlighting the significance of apprehending the dynamics of water content in plant tissues when cooling and heating as part of the development of a dependable cryopreservation protocol. Although the study recommends the standard DSC method for day-to-day work with known thermal properties of the sample and non-overlapping thermal events, QITTMDSC is endorsed for exact measurement of heat capacity in equilibrated conditions, which can help in identifying the state of matter.

This book is the contribution of more than thirty-five authors from eighteen institutions in seven countries. A rich source of information is provided on current developments and reviews of previous research as well as existing standard operating procedures. The book brings to the fore that although cryopreservation has progressed very well, there is much more to be explored, particularly in the case of CPAs. I am personally interested in having CPAs that will not compromise the biology of cells before they are exposed to sub-zero temperatures and will allow tissues to function normally after thawing. In a situation where there are no restrictions to the quantity of CPAs that can be used, the subjection of cells to cryopreservation will be effortlessly conducted. Though the question still persists: "Is cryopreservation just an illusion for some biological systems?" Let us work towards answering this question as we pursue cryobiology and cryopreservation.

Dr. Marian Quain
Council for Scientific and Industrial Research,
Ghana

Section 1

Recceing Oligomers, Polymers and Other Cryoprotective Agents

Chapter 1

The Use of Chitooligosaccharides in Cryopreservation: Discussion of Concept and First Answers from DSC Thermal Analysis

Hugo Desnos, Pierre Bruyère, Magda Teixeira, Loris Commin, Gérard Louis, Stephane Trombotto, Amani Moussa, Laurent David, Samuel Buff and Anne Baudot

Abstract

The use of dimethyl sulfoxide (Me_2SO) as a cryoprotectant agent (CPA) is controversial. Indeed, this cryoprotectant agent (CPA) is cytotoxic and potentially mutagenic. Therefore, other cryoprotectants must be used to reduce the proportion of Me_2SO in slow-freezing solutions. In this chapter, we propose to present the first evaluation of new non-penetrating cryoprotectants: the chitooligosaccharides (COS). These molecules are chitosan oligomers, which are biocompatible, antioxidant, and bacteriostatic. We first review the use of saccharides through cryopreservation processes. We question the possibility to reduce penetrating CPA during slow-freezing procedures. We propose to use COS as extracellular CPA to reduce the use of Me_2SO. We question the biocompatibility of COS on mouse embryos through the analysis of the cells' development. Next, we evaluate these molecules in slow-freezing solutions with a reduced quantity of Me_2SO. Our experimental approach is a physical method often used to characterize slow-freezing solutions. Differential scanning calorimetry (DSC) allows to evaluate the crystallization and melting processes, the amount of crystallized water, and the equilibrium temperature and consequently to evaluate the impact of different cryoprotectants. This study gives a better understanding on how slow-freezing protocols could be improved with extracellular CPA.

Keywords: slow-freezing improvement, chitooligosaccharides, dimethyl sulfoxide, physical approach, differential scanning calorimetry

1. Introduction

Despite the progress achieved these last decades, the improvement of cryopreservation procedures is still desired by the scientific community [1, 2]. A possible improvement is the decrease of the penetrating cryoprotective agents (CPA). The small and osmotically active penetrating molecules used as penetrating CPA in cryopreservation present risks of cytotoxicity for cells [3–6]. These molecules penetrate easily into tissues and, after impregnation, are difficult to fully extract from biological systems. Among the different penetrating CPA, Me_2SO is a penetrating CPA especially efficient to promote successful cryopreservation. However, the Me_2SO molecule presents risks of toxicity to biological materials [7–14]. It is also suspected to be an intercalator of DNA [15]. In this chapter, we propose to study the possibility to reduce the use of the Me_2SO molecule while maintaining the survival rate of cells in slow-freezing cryopreservation.

A challenge of modern cell cryopreservation is to propose procedures in which CPA concentration is as little as possible:

- In vitrification procedures, the techniques seek to obtain the glassy state with very small amounts of penetrating CPA (less than 25% (w/w)) by replacing them with non-penetrating CPA [3, 16–18]. Accordingly, the cooling must be applied as quickly as possible so that the cells do not suffer the damage associated with excessive dehydration and/or volume loss at room temperature. They also require the use of particularly small sample volumes (<1 μL) to reach high cooling rate that will prevent ice crystallization. These procedures are particularly delicate and require a certain dexterity of the experimenter [19, 20].

- In slow-freezing procedures, reducing the initial amount of penetrating CPA significantly increases the stresses and risks experienced by cells during cryopreservation. It is therefore difficult to propose an excessive reduction in the initial concentration of penetrating CPA. Indeed, it has been theoretically demonstrated, through mathematical modeling, that it is impossible to cryopreserve cells with initial Me_2SO concentrations below 1 $mol.L^{-1}$ [21].

We assume, however, that a good combination of penetrating and non-penetrating CPA and an adaptation of the slow-freezing protocols could provide satisfactory survival rates, despite a reduction in the initial proportion of Me_2SO used. This chapter investigates an alternative to the current slow-freezing procedures using Me_2SO and proposes procedures where the initial and necessary amount of Me_2SO is lowered.

Decreasing the Me_2SO proportion in solution will modify the couple "protocol/solution" within the cryopreservation procedure since it is necessary to counterbalance the loss of the cryoprotective effect of the removed Me_2SO molecules. To that end, this chapter studies the possibility to use chitooligosaccharide (COS) compounds in solution. The role of extracellular CPA is discussed, and the effect of the COS in solution is thermodynamically evaluated.

Even if common characteristics of various CPA can be defined, the substances currently used as CPA have different chemical structures and sizes, so it is still difficult to predict the cryoprotective properties of substances from their basic chemical structures. The modes of action of cryoprotectants have not yet been fully elucidated. Thus, the development of effective cryopreservation solutions remains primarily based on empirical considerations. Consequently, studying the potential cryoprotective characteristics of new compounds serves a dual purpose: (1) propose new cryoprotectants and (2) understand the mechanisms of cryoprotection. It makes possible to better classify the molecules according to their modes of action regarding cellular cryoprotection.

This chapter aims to assess the use of COS to fulfill part of the cryoprotective role of the Me_2SO. It presents first, from the cellular cryopreservation point of view, the interesting properties of saccharides (Part 2). It discusses next the optimization of the procedures required to replace a part of the penetrating CPA by using non-penetrating saccharides (Part 3). The use of COS to realize this optimization is explained (Part 4). Then, COS synthesis is presented from a physicochemical point of view (Part 5). The biocompatibility of COS on mouse embryos is questioned (Part 6), and a first understanding of the action of COS as cryoprotectants is quantified by DSC analysis (Part 7), before a conclusion on the possibility of using these compounds as non-penetrating CPA.

2. State of the art: the use of saccharides in cryopreservation

The cryoprotective role of saccharide compounds (mono-, di-, tri-, oligo-, or polysaccharides) is known for a long time. Some of them are secreted by certain cold-blooded organisms that withstand harsh winters [22, 23]. Several saccharides are water soluble, have a high osmotic effect, and have many influences on the solutions' properties: they can bind part of the water and make it non-crystallizable [24–29]; their action on viscosity [30, 31] has been highlighted, as well as their ability to disrupt the organization of water molecules in the liquid phase [32–35]. In a mixture of "H_2O/saccharide compound," the proportion of water has an influence on the properties of the saccharide compound and therefore on the properties of the mixture [29, 36].

These compounds are known to be unable to penetrate passively through most cell membranes. From the cryoprotective point of view, they act outside the cells, giving them limited action on cell protection. This is the reason why the standard in cryopreservation procedures remains the massive use of penetrating CPA that possess a high osmotic effect in both extra- and intracellular solutions. However, the beneficial use of saccharides to substitute penetrating CPA compounds has been proven in recent years. Indeed, saccharides have been used for cell cryopreservation in slow-freezing [37–51] or vitrification procedures [16–18, 39, 52–55]. A non-exhaustive overview from the literature of the use of the saccharides is displayed in **Table 1**.

Using empirical methods, it has been shown that these compounds allow the improvement of the post-cryopreservation survival rates of certain cell types [3, 37, 38, 44–48, 82]. Some authors have discussed the benefits of using them:

- Meryman was the first to propose that the presence of large polymers in solutions prevents the risks of osmotic shocks (when the penetrating CPA are released from the intracellular medium) as well as the denaturation of cell membranes [83].

- In cryopreservation of spermatozoa, small saccharides are often used as substituents for penetrating CPA [42, 49, 50, 61]. The advanced argument of their use is that they increase the viscosity, increase the glass-forming tendency, and protect the cellular lipid membranes [42, 61, 82].

Monosaccharides	Disaccharides	Trisaccharide
Glucose [56–63]	Sucrose [61, 64–67]	Raffinose [14]
Fructose [14]	Maltose [35]	**Polysaccharides**
Sorbitol [14]	Trehalose [16, 35, 39–43, 49–51, 61, 68–79]	Dextran [1, 2, 17, 18, 54, 67, 80, 81]
Mannitol [14]		
Galactose [64]		Ficoll [1, 17, 18, 54, 64, 66, 81]

Table 1.
Non-exhaustive directory from literature of saccharides used in cryopreservation procedures.

- In cryopreservation by vitrification, Kuleshova et al. [17] proposed to replace a large portion of penetrating CPA by large extracellular polysaccharides (Ficoll and dextran).

- In slow-freezing cryopreservation, saccharides have also been used [84]. For the cryopreservation of human or mouse embryos, Dumoulin et al. [81] highlighted the ability of large molecules (e.g., dextran) to reduce damage to zona pellucida when added in small amounts (i.e., less than 10% (w/v)). For the cryopreservation of cat embryos, Gómez et al. [67] were able to obtain interesting survival rates (>80%) by combining propylene glycol (1.4 M), sucrose (0.125 M), and dextran 10% (w/w). Slow-freezing procedures for some weak cellular systems have been proposed, without recourse to CPA penetrants [43]. This strategy is based on the induction, prior to ice formation, of cellular dehydration using osmotically active extracellular compounds (e.g., trehalose).

- For the slow-freezing cryopreservation of human red blood cells, Bailey et al. [85] reported the possibility to reduce the use of Me_2SO from 10% (w/w) to 2.5% (w/w) through the use of high concentration of copolymers (polyampholytes). These authors concluded that because no ice recrystallization inhibition (IRI) activity could be seen, the interaction of those copolymers with cells' membranes should be the reason that increases the cell recovery.

From these readings, mechanisms can be advanced to explain how the addition of saccharides in solution improves cell survival:

- Saccharides have a cryoscopic influence in the extracellular medium [86]. It reduces the damage associated with the formation and growth of the crystalline phase.

- Saccharides promote cell dehydration on cooling [17]. This aspect is ambivalent because, on the one hand, it allows to promote intracellular vitrification, but on the other hand, it increases the risks associated with cellular dehydration.

- Saccharides reduce the cellular rehydration during warming [19, 83]. By having an osmotic effect from the extracellular medium, they limit the osmotic gap (appeared during cooling, due to the difference between the intra- and the extracellular vitrification temperature [87]) between intra- and extracellular media. During warming and during the ice melting, saccharides guarantee a smoother return of water into the cells [19, 83].

- Saccharides participate in the evolution of extracellular viscosity. This reduces the kinetics of crystal growth and favors the deviation of the system from its equilibrium position. This effect has the consequence of improving the extracellular vitreous state achievement [17].

- Saccharides allow to "encapsulate" cells. Thus, the presence of large molecules in solution protects the zona pellucida of embryos during slow-freezing procedures [81]. Likewise, the presence of these large molecules can move the cells away from the ice in the overconcentrated amorphous phase [82]. It has been also reported that the presence of mono- or disaccharides in solution improves sperm motility after cryopreservation [88]. Finally, it has been proposed to create hydrogel-based microspheres to encapsulate cellular systems in small volumes [89].

- Saccharides promote the lipid transition of phospholipids from cell membranes. By interacting with cell membranes, saccharides have an effect on the fluidity of lipidic membranes [82, 83]. Interactions between membrane phospholipids and saccharides in solution have also been advanced as a method of promoting the success of cryopreservation procedures [90, 91].

Saccharidic CPA are used in different ways depending on their physicochemical characteristics. In general, the polysaccharides are added in solution to increase the viscosity while having a weak osmotic activity. On the contrary, the small saccharide molecules (such as glucose) are added in solution to have high osmotic activity within the solution (this favors and induces the cellular dehydration), without changing the viscosity too much.

3. Strategy for improving slow-freezing cryopreservation procedures

3.1 Foundations of the strategy

Models showed that the improvement of cryopreservation procedures is impossible if it implies an excessive reduction in the use of penetrating CPA (<1 mol.L^{-1}) [21]. But these models studied this reduction without trying to use other CPA compounds with completely different properties than those currently used. Me_2SO is certainly one of the best compounds to fulfill the roles that are required to a CPA (apart from its cytotoxicity), but it is conceivable that a combination of different CPA may also be compatible with cell survival.

To reduce the use of Me_2SO in slow-freezing cryopreservation procedures, several assumptions were considered, based on results published in the literature:

a. The slow-freezing procedures have been optimized according to some parameters, but new optimizations are possible.

Other combinations of the parameters may be better than the currently used. The search for new CPA is still ongoing, and the cryopreservation procedures are better understood. The discovery of intracellular vitreous transitions [87], the control of the IRI process [92], the development of new cooling or warming techniques [93], etc. may suggest that optimization of the procedure is still possible.

b. The optimization of the proportion of penetrating CPA in the cryopreservation solution is considered in order to find a formulation offering good vitrification conditions for the remaining solution.

It has been shown that the proportion of penetrating CPA necessary to lead to a cellular vitrification can be reduced if there is a high cellular dehydration [16]. Indeed, the intracellular cytoplasmic medium has a great glass-forming tendency because of the presence of many macromolecules [87, 94–96]. The presence in intracellular solution of a large proportion of proteins and other organic compounds (macromolecules), accompanied by dehydration of the medium, is sufficient to promote vitrification [87, 94]. In addition, experiments have shown the existence of colloidal glass transitions in cells during their dehydration [96, 97]. These transitions can promote intracellular vitrification and potentially protect cells (since cell cytoplasm is often heavily loaded with large molecules, it can easily vitrify [17, 59, 96]). In a vitrification procedure, using successive baths of osmotic equilibration, the cells are placed in solutions containing penetrating CPA. It was proposed by Kuleshova et al. [17] to replace, in those procedures, a large proportion of penetrating CPA with non-penetrating polymers (PVP, dextran, Ficoll, etc.). In the last osmotic equilibration bath, non-penetrating CPA (sugars or polymers) are added to promote cell dehydration just before the plunge into liquid nitrogen (LN2). The osmotic effect of these non-penetrating CPA promotes the vitrification of the intracellular medium and ensures the success of the procedure with a lower amount of penetrating CPA [16]. By adaptation to slow-freezing procedures, we assume that it is possible to force the intracellular colloidal glass transition to the correct temperature by suitably combining temperature lowering and cellular dehydration. However, cellular dehydration is not possible over long periods when the temperature remains high. During a vitrification procedure, the dehydration time is short (primarily applied during the last osmotic bath, a few seconds before the plunge into LN2). During slow-freezing, if the dehydration intervenes too extremely, it could lead to cellular collapse [83, 98, 99]. To solve this problem, we rely on the work of Mazur et al. [100] who proposed an alternative to conventional cellular cryopreservation protocols by rapidly bringing the cells to a temperature low enough to make the effects of cell dehydration less deleterious. This temperature must, however, remain above the nucleation temperature of the intracellular medium. A temperature stabilization is then carried out at this intermediate temperature to allow the equilibration of the media. Then, the cooling continues at a slow cooling rate until the plunge into LN2. This type of protocol is called "rapid cooling interrupted" [100]. It should be noted that this protocol also possesses a seeding step in order to control the extracellular ice growth. We assume that, when using this type of protocol, it would be possible to allow faster cell dehydration already biologically tolerated in the case of vitrification procedures.

c. The use of non-penetrating CPA can allow reducing the necessary proportion of penetrating CPA by optimizing the procedure of cooling and warming.

For a slow-freezing procedure, in which ice formation is allowed and desired, the osmotic effect of a non-penetrating CPA becomes increasingly important as the proportion of water in the remaining extracellular medium decreases. Conversely, in the intracellular medium, it is mainly the penetrating CPA that participates in the osmotic balance. There is therefore a gap between these two media in the osmolality evolution, which accentuates cellular dehydration and increases cell contraction at a given cooling rate. The use of an osmotically active non-penetrating CPA in solution may induce a significant change in the osmolality difference between the intra- and the extracellular media.

3.2 The elements of the proposed optimization

The toxicity of the cryopreservation solutions is related to the CPA concentration reached in the remaining solution to enable the vitrification of the intracellular medium. Moreover, to protect the cells, the vitrification must also be carried out in the noncrystalline extracellular medium in contact with the cells. Intracellular vitrification is dependent on the presence of penetrating CPA and on the cellular dehydration. In consequence, the intracellular vitrification is dependent on the diffusion of materials across the cellular membrane. That is why cryopreservation procedures are time dependent. Indeed, an ideal cooling rate exists at which, for a specific cell (i.e., a specific permeability and cell size), the cellular dehydration is ideally compensated by the CPA penetration. This rate deals with the two major risks of the slow-freezing procedures, the "solute effect," which occurs at low cooling rate, and the intracellular nucleation, which occurs at high cooling rate [101].

The addition of penetrating CPA in a solution, in a reasonable proportion (> 1 mol.L^{-1} for slow-freezing and > 3 mol.L^{-1} for vitrification), is considered essential for the survival of mammalian embryo type cells [21]. Very few studies have been published on attempts to reduce the amount of Me_2SO in solution during slow-freezing cryopreservation [43, 84]. This strategy was evaluated by relying on modeling arguments, showing that it seems complicated to reduce this initial proportion of penetrating CPA [21]. We make a proposal for slow-freezing procedure optimization using a reduced initial proportion of penetrating CPA, supplemented with non-penetrating CPA.

The decrease of the initial proportion of penetrating CPA implies an increase in the amount of crystallizable water, a stronger evolution of the crystal/liquid ratio in the solution with temperature, a higher amount of penetrating CPA needed to diffuse through the membrane, more risks of contact between cells and ice crystals, etc. Thus, penetrating CPA play a cryoprotective role that cannot be fully reproduced in the extracellular environment by non-penetrating CPA. Moreover, the stresses and risks associated with the ice formation must be limited or prevented until the plunge to LN_2. It is therefore necessary to continue to use penetrating CPA, but their amount in the initial solution should be limited to reduce cytotoxicity. In addition to the presence of penetrating CPA, the vitrification of the extracellular remaining solution depends on the presence of non-penetrating CPA that limit the evolution of penetrating CPA concentrations and reduce the risks associated with the ice formation in solution. The reduction of penetrating CPA implies to control the damages related to the ice formation in solution by replacing a large part of the initial extracellular water by non-crystallizable substances, which offer favorable conditions for cryopreservation. However, the biocompatibility of these compounds on the cellular systems must be guaranteed.

Because the proportion of CPA in the intracellular medium has a lighter impact than the cellular dehydration on the intracellular vitrification [16, 87, 94–96], we assume that the proportion of penetrating CPA may be limited. In that case, it is then necessary to reduce the deleterious effects related to the cellular dehydration by using non-penetrating CPA that are not highly osmotically active and whose molar concentrations slightly change during cooling (i.e., compounds with a large molar mass). It is also necessary to employ the "fast interrupt protocol" by suitably combining dehydration and lowering temperature so that the colloidal intracellular transition occurs before cell dehydration becomes too important.

Finally, it is necessary to verify that the extracellular remaining solution can vitrify easily so that the cryopreservation procedure can be interrupted at a higher plunge temperature into LN_2. Thus, the use of cryostabilizer CPA that facilitate the achievement of the vitreous state, without binding a significant amount of water, may promote the stabilization of the system "ice/remaining solution" (by reducing the risk of recrystallization of the intercrystalline remaining solution).

We chose to study the possibility to cryopreserve mouse embryos with 5% (v/v) of Me_2SO in the initial solution, supplemented with non-penetrating CPA (COS).

4. The proposition to use COS

The idea of using COS compounds in cell cryopreservation is supported by:

- An absence of toxicity, an ability to degrade without toxic residues, and a biocompatibility [102].

- Bacteriological, fungistatic, and antitumor properties [102–106].
- An ability to be soluble in aqueous solutions at physiological pH [107].
- A tendency to form aggregates in aqueous media [108].
- A chemical structure of oligosaccharide type close to molecules currently used in cryobiology.
- Potential interactions with water and consequently a potential effect on the properties of aqueous solutions (like that proposed by mono-, di-, and oligosaccharides).

In addition, because of the length of their chains, the COS propose intermediate properties between the polymers and the mono- or disaccharides. These properties are adjustable since the length of the chains can be selected during their synthesis using degree of polymerization (DP) parameter as well as the nature of the monomers present on the chain using the degree of acetylation (DA) parameter. These modular characteristics let us hope for the attainment of physicochemical properties that are adjustable for use as a CPA. Furthermore, they are particularly biocompatible and nontoxic. As a result, addition of COS to some penetrating CPA appears to be a relevant choice to reduce the needed amount of CPA during slow-freezing.

Assuming that a use of COS would reduce the use of Me_2SO, this work proposes to look for a composition of COS-based solution that can reproduce, in the extracellular medium, some effects of Me_2SO (favor the vitrification of the intracellular medium, promote the vitrification of the intercrystalline remaining solution, reduce the cell dehydration, etc.).

Our current procedure used for mouse embryo cryopreservation has been optimized for the "IMV" holding medium (embryo-holding medium, IMV® Technologies, L'Aigle, France) supplemented with approximately 10% (v/v) Me_2SO [109]. We set the goal of reducing by 50% the proportion of Me_2SO, to successfully cryopreserve mouse embryos using an IMV + 5% (v/v) Me_2SO solution containing a certain amount of COS.

5. Synthesis and chemical characterization of COS

The basic compound used to synthesize COS is chitosan supplied by the Indian company Mahtani Chitosan®. This chitosan (batch 244/020208) is produced from chitin extracted from shrimp shells. The chitosan provided was almost completely deacetylated (DA < 0.5%) by deacetylation reaction. This chitosan has a number-average molecular weight of 115 $kg.mol^{-1}$ and dispersity of 2.3. For the preparation of COS, the macromolecular chitosan chains are depolymerized according to a nitrous acid deamination reaction using sodium nitrite ($NaNO_2$) in acidic conditions [110]. It is only possible to obtain a statistical average chain length distribution around a mean DP value. To control the DA parameter, it is possible to perform an N-acetylation reaction of the D-glucosamine units with acetic anhydride as acetylation agent. In this chapter, DP and DA parameters have been used to name the COS as follows: COSDP_DA, with DP = "the average number of monomers per COS chain" and DA = "the average degree of acetylation of the chains of this COS."

To estimate the residual hydration rate in purified COS, we used data from the literature. A previous study of the same types of chitosan chains, but in a polymeric form, evaluated these hydration levels [111]. This study performed by thermogravimetric analysis has shown that there are residual hydration levels after synthesis, purification, and lyophilization, which are variable according to the DA parameter of the chitosan polymers chains. From this data, we extrapolated the hydration levels of the COS compounds according to the DA parameter. For a DA $\approx$ 0%, we considered a residual hydration value of the COS compound equal to 6.5% (w/w). This is consistent with other results published in the literature where a value of 6% (w/w) was obtained [112].

6. Questioning the biocompatibility of COS

Before considering the use of COS in cryopreservation solutions, an evaluation of its biocompatibility is necessary. It ensures that their presence will not be deleterious to the future survival and development of the cryopreserved biological system. By their nature and chemical composition, the biocompatibility of COS has already been emphasized [113]. However, the molecular interactions between all the compounds in extracellular solution and the cell membranes are complex. There are, for example, interactions between COS molecules and cell membranes that are invoked to explain their fungistatic and bacteriostatic properties [113]. It is therefore not excluded that interactions may exist with eukaryotic cells. In addition, the cell system chosen for this study, the mouse embryo, is a fragile system whose development can be disrupted by the presence of harmful materials. Consequently, deleterious influence on this cellular system, linked to COS, must be excluded.

6.1 Materials and methods

The biocompatibility of COS was evaluated with mouse embryos. A procedure identical to that presented in a previous team article [109] was applied for the production of embryos used in this study. To study the biological action of COS on mouse embryos, a large quantity of COS10_0 (150 $mg.mL^{-1}$) was dissolved in the cryopreservation solution (IMV + 5% (v/v) Me_2SO) and in the "IMV" holding medium. One hundred twenty-six embryos (morula stage) were collected and were then mixed and divided into five groups (cf. **Table 2**): (1) "IMV," a control group placed in a solution without CPA; (2) "IMV + 10% (v/v) Me_2SO," a control group placed in the solution conventionally used for cryopreservation of mouse embryos; (3) "IMV + 5% (v/v) Me_2SO," a control group placed in the solution in which COS

	Groups	pH	Number of embryos in culture	Young blastocysts		Expanded blastocyst		Hatching	
1	IMV	7.29	25	23	92%	23	92%	23	92%
2	IMV + 10% (v/v) Me_2SO	7.29	25	20	80%	18	72%	14	56%
3	IMV + 5% (v/v) Me_2SO	7.30	25	23	92%	20	80%	20	80%
4	IMV + 5% (v/v) Me_2SO + (150 $mg.mL^{-1}$) COS_{10_0}	7.32	25	25	100%	22	88%	20	80%
5	IMV + (150 $mg.mL^{-1}$) COS_{10_0}	7.34	26	24	92%	23	88%	23	88%

The stage reached by the embryos was evaluated under a binocular microscope every 24 hours for 4 days. For each embryonic stage, the number of embryos observed at this stage is indicated in the left column and, on the right, the corresponding percentage in relation to the initial number of cultured embryos.

Table 2.
Compilation of embryonic developments observed by groups.

are diluted; (4) "IMV + 5% (v/v) Me_2SO + (150 mg.mL^{-1}) COS_{10_0}," a test group placed in a tested cryopreservation solution with COS; and (5) "IMV + (150 mg. mL^{-1}) COS_{10_0}," a second test group, without Me_2SO, studied to highlight the effect of COS and avoid the potentially cross-cutting effects between COS and Me_2SO.

The embryos were placed in these solutions at room temperature for 10 minutes. They were rinsed with M16 culture medium (IMV® Technologies, L'Aigle, France) and then placed in a culture chamber. The culture medium was equilibrated in the incubator (+37°C; 5% (v/v) CO_2; humid atmosphere), and then the embryos were introduced therein. The culture was maintained and supervised for 4 days. The development of embryos was then compared (every 24 hours) with other control groups to assess the state of the embryonic stage reached by each embryo. These comparisons can highlight certain harmful effects related to the presence of the different products on cellular development.

6.2 Results and discussions

After 24 hours, 20% of the embryos in the 10% (v/v) Me_2SO group failed to reach the "young blastocyst" stage. For groups with 5% (v/v) Me_2SO, the development stage remains similar to the solution without Me_2SO. The presence of COS does not appear to have a deleterious effect on the development of embryos at this stage, and the reduction in the proportion of Me_2SO in solution seems to be beneficial.

After 48 hours, it is possible to compare the number of embryos which reached the "expanded blastocyst" stage to the number of embryos which previously reached the "young blastocyst" stage. The numbers are the same for group 1 (control without CPA and COS), while an embryo did not develop in group 5 (with COS but without CPA), and more than one embryo did not develop in the other groups (2, 3, and 4). By comparing the results for groups 3 and 4, the effect of COS presence on embryo development may be considered minimal. Conversely, the decrease in the number of living embryos in groups 2, 3, and 4 seems to be directly associated with the presence of Me_2SO in the solution where embryos are bathed at room temperature.

After 72 hours, the number of embryos that reached the hatching stage is the least important for group 2 (with 10% (v/v) Me_2SO). There is no difference according to the presence or absence of COS in the IMV, and the final percentage in each case is very close. It is the same in solutions with 5% (v/v) Me_2SO, where the number of embryos which reached this stage of development is equivalent, with or without COS.

According to this study repeated only once, a negative effect of Me_2SO on the development of mouse embryos is highlighted. However, a negative effect of COS on this cellular development is discarded. These results should be repeated but seem to demonstrate the biocompatibility of COS for the mouse embryos. We conclude that COS can be used in the cryopreservation solutions in contact with cells.

7. Thermodynamic characterization of cryopreservation solutions containing COS

The strategy outlined in Part 3 aims to reduce by 50% the initial volume proportion of Me_2SO in slow-freezing solutions with the help of COS. It is thus necessary to know more precisely their mode of action in aqueous solutions. Several studies have already proposed the thermodynamic characterization of chitosan or chitosan derivatives [26, 112, 114–116]. However, these studies have focused on the thermodynamic characterization of polymer chains. To our knowledge, the characterization of the thermodynamic properties of aqueous solutions formulated with soluble oligosaccharide derivatives of chitin or chitosan has not yet been done.

The objective of this study is to evaluate the impact of COS on the thermodynamic properties of aqueous solutions and to put forward a potential action of these products on water and other properties of the solutions (viscosity, gelling, etc.). It will help to evaluate the cryoprotection effect of COS.

7.1 Materials and methods

7.1.1 Thermodynamic characterization of solutions

A differential scanning calorimeter (DSC Diamond; Perkin Elmer) with power compensation was used in this study. The previously described procedures [117] have been applied, with the same methods of sample preparation, calibration of the DSC calorimeter, and precautions for use. COS were added in the IMV + 5% (v/v) Me_2SO solution, with different mass concentrations ($mg.mL^{-1}$) of hydrated powder of COS (per unit volume of the IMV + 5% (v/v) Me_2SO solution), as listed in **Table 3**.

To study the cryoprotective capability of COS as a substitute for Me_2SO, the solutions were studied by DSC using a protocol described elsewhere [117]. This protocol starts with a cooling from +10°C to −150°C at $-100°C.min^{-1}$, a warming from −150°C to +10°C at $+2.5°C.min^{-1}$, a cooling from +10°C to −150°C at $-2.5°C.min^{-1}$, and a warming from −150°C to +20°C at $+20°C.min^{-1}$. The experiments were repeated three times (n = 3) for each concentration.

7.1.2 Correction of the studied mass concentrations

The studied COS have a residual moisture content. Thus, the mass concentration of the added powder is not equivalent to the real mass concentration of the added product. Correction is made according to Eq. (1):

$$\textit{Real mass concentration of COS} = \textit{Powder mass concentration of hydrated COS} * \left(1 - \Psi_{H_2O,\,ini}(COS)\right) \quad (1)$$

Real mass concentration of COS ($mg.mL^{-1}$), powder mass concentration of hydrated COS ($mg.mL^{-1}$), $\Psi_{H_2O,\,ini}(COS)$, residual hydration ratio in the lyophilized powder of purified COS (Ø)

Products	Averaged DP	Averaged DA (%)	Residual hydration ratio after purification % (w/w)	Powder mass concentrations studied ($mg.mL^{-1}$)	Real mass concentrations studied ($mg.mL^{-1}$)	DSC sample mass (mg)
$COS_{7.5\_0}$	7.5	0	6.5	30; 60; 90; 150	28.05; 56.1; 84.15; 140.3	~5
COS_{10_0}	10	0	6.5	0; 25; 50; 100; 150; 200	0; 23.38; 46.75; 93.5; 140.3; 187	~10

Table 3.
Table of the studied products, with their characteristics (DP, DA, and residual hydration ratio after purification) and their studied mass concentrations.

The actual mass concentrations of COS in the solutions that were studied are presented in **Table 3**. In the following analyses, only the actual mass concentration of COS is considered for a given compound. However, to simplify the notation in thermograms, the hydrated powdered mass concentration of COS is used to designate the solutions.

7.1.3 Normalization of the transition enthalpies to 0°C

The amounts of ice formed in solution were estimated using the determination of the crystallization enthalpy (ΔH_c) and the melting enthalpy (ΔH_m). To estimate these amounts, the peak areas were calculated using a sigmoidal curve baseline (with the Pyris software 11.1.1). During warming the tangents, allowing the calculation of the sigmoidal line, were positioning, respectively, just before the colloidal transition and just after the melting peak.

In order to compare the transition enthalpies, a normalization of the measured values to a specified temperature of 0°C was made. As the enthalpy value of the transition of water to ice evolves as a function of temperature, the "latent heat of solidification of supercooled water" (Lf) was obtained at the measured phase change temperature from Boutron's [118] data interpolation. His data were calculated [119] from Angell's [120] specific heat capacity (Cp) of supercooled water and from Weast's Hand Book's [121] Cp of ice at different temperatures.

The normalization of the measured values to the expected values at the specified temperature of 0°C considers that the same proportion of solution transits at the transition temperature (T_t) ($\Delta H_{meas}(T_t)/L_f(T_t)$) and at 0°C ($\Delta H_{normalized}(0°C)/L_f(0°C)$). This equality leads to Eq. (2):

$$\Delta H_{normalized}(0°C) = \frac{\Delta H_{meas}(T_t) * L_f(0°C)}{L_f(T_t)} \tag{2}$$

where $\Delta H_{normalized}(0°C)$ is the transition enthalpy of the sample normalized to 0°C ($J.g^{-1}$), $\Delta H_{meas}(T_t)$ is the transition enthalpy of the sample measured at a transition temperature T_t ($J.g^{-1}$), $L_f(T_t)$ is the transition enthalpy of a pure water sample measured at a transition temperature T_t ($J.g^{-1}$), and $L_f(0°C)$ is the transition enthalpy of a pure water sample measured at 0°C (= 333.4 $J.g^{-1}$) ($J.g^{-1}$).

During cooling, to analyze the crystallization enthalpies, the nucleation temperature (T_n) was considered as the transition temperature (i.e., $T_t = T_n$). During warming, to analyze the melting enthalpies, the temperature of the summit of the melting peak (T_{max}) was considered as the transition temperature (i.e., $T_t = T_{max}$). These are approximations because for a mixture there is not only one transition temperature.

7.1.4 Comparison of measured and calculated transition enthalpies with the addition of an inert mass

When a non-hydrated inert mass is added in a solution, a lowering of the enthalpies of crystallization and melting is expected since this addition decreases, in proportion, the mass of water present in solution. The term "inert" refers to a product that does not influence the transition enthalpies (i.e., a product that does not transit and does not influence the transition of other products). The theoretical link between the decrease of the transition enthalpies and the decrease of the mass proportion of water in solution is given by Eq. (3). The comparison of a theoretical value with a measured one allows estimating the mass of water which transits in comparison to the mass of water present within the solution. Experimental values lower than the theoretical ones materialize an evolution of the proportion of solution which is crystallized. Indeed, it materializes a deviation in the evolution of the compound concentrations within the remaining solution (according to the lever rule).

$$\Delta H_{\text{calculated with an inert mass}}([c]) = \Delta H_{\text{measured}}([c] = 0) * \Psi_{\text{IMV+5\%(v/v)Me}_2\text{SO}} \tag{3}$$

where $\Delta H_{\text{calculated with an inert mass}}([c])$ is the calculated specific enthalpy for the transition of liquid water to ice for an IMV + 5% (v/v) Me_2SO solution with a concentration [c] of inert mass ($J.g^{-1}$), $\Delta H_{\text{measured}}([c] = 0)$ is the experimentally measured specific enthalpy for the transition of liquid water to ice for a IMV + 5% (v/v) Me_2SO solution = 265.68 ± 3.2 $J.g^{-1}$ (this specific enthalpy value corresponds to the average of the three measurements with the IMV + 5% (v/v) Me_2SO solution without COS) ($J.g^{-1}$), and $\Psi_{\text{IMV+5\%(v/v) Me}_2\text{SO}}$ is the initial mass proportion of the IMV + 5% (v/v) Me_2SO solution in the prepared IMV + 5% (v/v) Me_2SO solution containing an inert mass (Ø).

Eq. (4) allows estimating the initial mass proportion of the IMV + 5% (v/v) Me_2SO solution in the prepared IMV + 5% (v/v) Me_2SO solution containing an inert mass such as COS.

$$\begin{aligned}\Psi_{\text{IMV+5\%(v/v)Me}_2\text{SO}}(\text{sol}) &= \frac{m_{\text{IMV+5\%Me}_2\text{SO}}}{m_{\text{IMV+5\%Me}_2\text{SO}} + m_{\text{product}}} \\ &= \frac{d_{\text{IMV+5\%Me}_2\text{SO}} * \rho_{H_2O} * V_{\text{sol}}}{d_{\text{IMV+5\%Me}_2\text{SO}} * \rho_{H_2O} * V_{\text{sol}} + [\text{product}] * V_{\text{sol}}} \\ &= \frac{1014.5}{1014.5 + [\text{product}]}\end{aligned} \tag{4}$$

where $\Psi_{\text{IMV+5\%(v/v) Me}_2\text{SO}}$ is the initial mass proportion of the IMV + 5% (v/v) Me_2SO solution in the prepared IMV + 5% (v/v) Me_2SO solution containing an inert mass (Ø), $d_{\text{IMV + 5\%Me2SO}}$ is the density of the IMV + 5%(v/v) Me_2SO solution = 1.0145 (experimentally measured with a pipette and a high sensibility weighing scale) (Ø), ρH_2O is the mass volume of water ≈ 1000 ($mg.mL^{-1}$), V_{sol} is the volume of the solution (mL), and [product] is the initial mass concentration of hydrated product added in solution ($mg.mL^{-1}$).

7.1.5 Estimation of the cryoscopic depression

A cryoscopic depression indicates the action of a dissolved product on the properties of water molecules in solution, providing information on interactions that occur between this product and the water molecules. The estimation of the maximal equilibrium temperature (T_m), for different concentrations of the same COS introduced in the IMV + 5% (v/v) Me_2SO solution, informs about the cryoscopic depression induced by the oligosaccharide. Moreover, knowing the temperature (T_m) of a solution allows to estimate, as a function of the temperature, the supercooling magnitude reached in this mixture (before crystallization).

Due to the kinetic phenomena of heat transfer, it is often imprecise to confuse the temperature recorded at the maximum of the endothermic melting peak (T_{max}) with the maximum melting point temperature of the ice in solution, because T_m is less or equal to T_{max} [119]. In the kinetic phenomena being influenced by the mass involved in the melting process, the comparison of the T_{max} values becomes difficult between the series for which the mass of sample studied is different and in which the mass of water in solution concerned by the melting is different.

Since the T_{max} values cannot be compared and can hardly be associated only with a cryoscopic phenomenon, we have not realized this comparison. We used an alternative method for estimating the cryoscopic effect based on the shape of the thermograms. It consists in the comparison of the size and the peaks spread with temperature [119].

7.2 Phase transition analysis of crystallizable water

The calculated values of the crystallization and melting enthalpies of the solutions were normalized to 0°C and represented in **Figure 1**. The calculation of the phase change enthalpy expected following the addition of an inert mass was performed in the range of the concentrations studied, using Eqs. (3) and (4). These values were plotted (**Figure 1**), with measured enthalpy values (normalized to 0°C), as a function of the initial COS concentration.

A sharp decrease in the ΔH_c and ΔH_m values is observed as a function of the mass concentration of COS introduced, and this decrease seems affine. For the same mass concentration of non-hydrated powder studied, the differences between ΔH_c and ΔH_m are small. Until 50 $mg.mL^{-1}$, a good correlation is observable between the calculated mass enthalpies following the addition of a non-hydrated inert mass and the measured mass enthalpies in the presence of COS (cf. **Figure 1**). COS molecules added in the solution can replace part of the water molecules present in the solution and thus reduce the amount of crystallizable water in mass proportion (as materialized by the orange line in **Figure 1**). To play an additional role, the CPA molecules must either bind to a portion of the water molecules to prevent their crystallization or promote the glass-forming tendency of free water during cooling. Based on **Figure 1**, COS do not seem to have significant effects, although these effects increase for the higher COS concentration (> 100 $mg.mL^{-1}$).

For the smallest COS concentrations, if the COS could bind to a part of the water, or highly favor the vitrification, then a difference between calculated enthalpies and experimental enthalpies or a difference between the crystallization enthalpies and the melting enthalpies (for a nominative mass of COS introduced in solution) would have been expected. These results imply that the addition of COS in

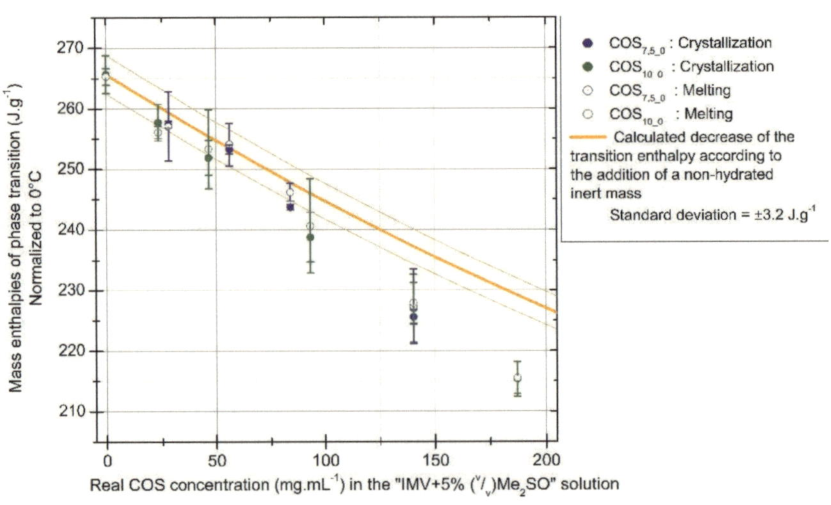

Figure 1.
Crystallization and melting enthalpies, normalized to 0°C, for the different solutions containing COS. Here the calculated transition enthalpies are also materialized in the case of the addition of a non-hydrated inert mass in solution. The standard deviations are calculated according to the three data obtained for each value. The standard deviation of the calculated enthalpy is the standard deviation of the measured value without COS.

DOI: http://dx.doi.org/10.5772/intechopen.89162

the solution does not disturb further the ability of water molecules to crystallize. Thus, the COS presence does not alter either the amount of bounded water or the quantity of bulk water that vitrified.

For the highest COS concentrations, the amount of water that can crystallize is reduced, as indicated by the smaller normalized values of ΔH_c and ΔH_m. This decrease in crystallization and melting enthalpies implies a decrease in the amount of ice formed. Because normalized values are lower than the calculated, we argue that it is the materialization of the capability of COS to stabilize the remaining solution and to limit the crystallization of a part of the water molecules in the remaining solution.

COS, therefore, serve the purpose of reducing ice formation during cryopreservation procedures. However, their mode of action and its consequences for cells deserve to be studied deeper.

7.3 Cryoscopic depression induced by the addition of COS

As could be seen in **Figure 2**, when COS is added, the maximum of the heat flow related to the melting endothermic peaks appears to be smaller, and the peaks are shifted toward the low temperatures.

When there is little ice in the sample, the effects responsible for the temperature difference between T_m and T_{max} decrease. Thus, the decrease in the amount of ice formed when the amount of COS increases (cf. **Figure 1**) could explain the lowering of T_{max} (cf. **Figure 2**) without cryoscopic effect. When there is a smaller amount of ice and fewer phenomena that may imply the shift of the temperature of the

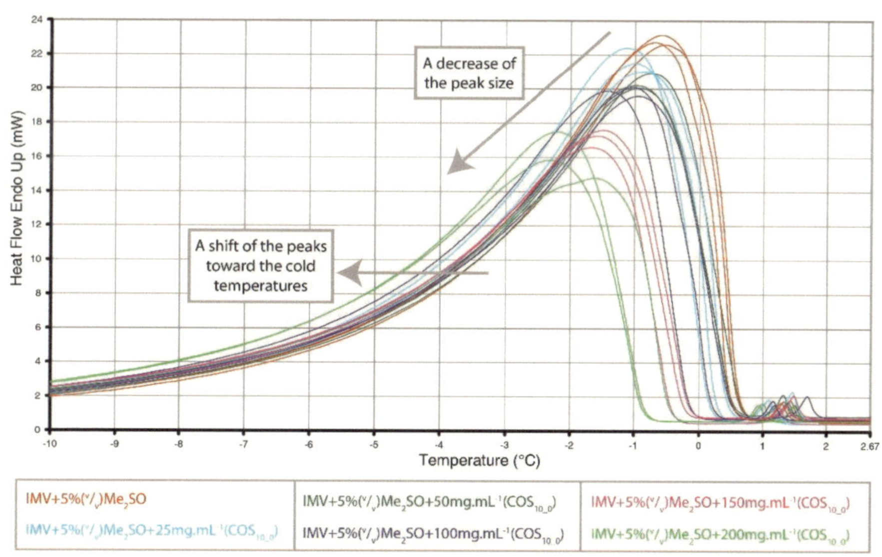

Figure 2.
Zoom on the bottom of the melting peaks obtained for the mixture: IMV + 5% (v/v) Me_2SO + [COS_{10_0}]. Observation of the COS influence on the shape of the melting peaks. Protocol, 2. Sample volume = 10 μL. Warming rate = +2.5°C.min^{-1}. Data range, normal. All cooling and warming thermograms were straightened up from −100 to −80°C and then lined up at 0.0 mW. The mass concentrations of the studied hydrated powder are equal to 0, 12.5, 25, 100, and 200 mg.mL^{-1}. Two endothermic peaks are observable on each of these thermograms, a large one which is associated to the melting of the ice previously crystallized within the sample and a small one which is an anomaly. This anomaly of melting is systematically present during the analysis of slow-freezing solutions, linked to the melting of the previously condensed and crystallized ambient humidity of the air encapsulated within the sample pan. This humidity could be from the previous ambient air in the laboratory or from the equilibration at the vapor pressure of water between the encapsulated air and the sample aqueous solution or from a combination of both.

melting peak summit, the melting peak is smaller, and then, the temperature of the peak's summit is shifted toward low temperatures.

However, the offset that is observed at the onset of the peaks cannot be associated with a reduction in the amount of ice that melts during warming. This shift indicates that there is a change in the amount of ice transiting at lower temperatures. The amount of water that melts as a function of temperature is greater at lower temperatures for the solutions with COS (cf. **Figure 2**), even if, in the presence of COS, the total amount of ice previously formed is lower. This phenomenon is the materialization of a slight cryoscopic effect linked to the addition of COS in solution. This effect is, however, difficult to quantify and remains below 0.5°C (cf. **Figure 2**).

The presence of a weak cryoscopic depression associated with the addition of COS in the IMV + 5% (v/v) Me_2SO solution involves the addition of COS having little or no influence on the osmolality of the studied solutions. Further verifications will nevertheless have to be made to verify these results, especially by studying the characteristics of the mixture: "H_2O-COS." Based on these preliminary findings, we will consider that COS compounds can be added to cryopreservation solutions without fear of a significant increase or acceleration of cell dehydration.

8. Conclusion on the COS usage for cryopreservation

According to this first thermodynamic characterization, the use of COS as CPA is debatable, as are their cryoprotective capabilities.

COS appear to be biocompatible in solution, and no disruption to mouse embryo development has occurred when COS were evaluated. COS have a significant influence on lowering the amount of crystallizable ice in solution, without participating in cell dehydration. Indeed, their addition in solution replaces a significant part of water, without significant osmotic effect. This is an interesting use in slow-freezing cryopreservation since it makes it possible to reduce the mechanical damage associated with the formation of ice, without change in the equilibrium of the osmotic balance with the intracellular media. We can thus assume that the use of such a product does not promote the mechanisms of dehydration of the intracellular medium.

However, COS do not seem able to influence the proportion of IMV + 5% (v/v) Me_2SO solution remaining liquid during freezing. As a result, they cannot participate in the main protections associated with CPA. Using these compounds without any other CPA has, therefore, to be discarded. COS do not participate in the equilibration of osmotic pressures between intra- and extracellular media. Thus, they dehydrate the overconcentrated residual solution, without modifying the mass proportion of crystallized water and without dehydrating the cells. They act only in the extracellular medium by inducing a steric hindrance and by increasing the solution viscosity. They seem to contribute to the stabilization of the extracellular noncrystalline medium.

Combined with penetrating cryoprotectants, we assume that COS can promote a decrease in the use of penetrating CPA while ensuring the successful vitrification of intercrystalline spans. Nevertheless, several questions remain about the action of COS in solution. It is particularly possible to question their interactions with water molecules. Similarly, it is difficult to know how these compounds interact with Me_2SO. In order to further characterize the role of these compounds in water solution, it will be necessary to characterize them in simpler binary solutions composed of COS and water. It would also be necessary to study the influence of COS on the ice crystal organization in cryopreserved samples (notably to study the influence of COS on the average size of intercrystalline spans). Further experiments, particularly on their ability to increase the glass-forming tendency of the remaining solution, should be conducted in the future.

Acknowledgements

This work was supported by a grant from the Veterinary Campus of Lyon (VetAgro Sup, Marcy l'Etoile, France) and the infrastructure project CRB-Anim (ANR11-INBS-0003).

Author details

Hugo Desnos[1,2*], Pierre Bruyère[1], Magda Teixeira[1], Loris Commin[1], Gérard Louis[2,3], Stephane Trombotto[4], Amani Moussa[4], Laurent David[4], Samuel Buff[1] and Anne Baudot[2,3]

1 Université de Lyon, VetAgro Sup, UPSP ICE 'Interactions Cellules Environnement' and CRB CryAnim (CRB-Anim), Marcy L'Etoile, France

2 Université de Paris, Campus Saint Germain des Près, Paris, France

3 Université de Paris, LVTS, Inserm U1148, Paris, France

4 Université de Lyon, IMP 'Ingénierie des Matériaux Polymères', CNRS UMR 5223, Villeurbanne, France

*Address all correspondence to: hugodesnos@hotmail.fr

References

[1] Shaw JM, Oranratnachai A, Trounson AO. Fundamental cryobiology of mammalian, oocytes and ovarian tissue. Theriogenology. 2000;**53**(1):59-72

[2] Pereira RM, Marques CC. Animal oocyte and embryo cryopreservation. Cell and Tissue Banking. 2008;**9**(4): 267-277

[3] Kasai M. Vitrificatio: Refined strategy for the cryopreservation of mammalian embryos. Journal of Mammalian Ova Research. 1997;**14**: 17-28

[4] Fahy GM. The relevance of cryoprotectant "toxicity" to cryobiology. Cryobiology [Internet]. 1986;**23**(1):1-13. Available from: http://linkinghub.elsevier.com/retrieve/pii/0011224086900131

[5] Fahy GM, Lilley TH, Linsdell H, Douglas MSJ, Meryman HT. Cryoprotectant toxicity and cryoprotectant toxicity reduction: In search of molecular mechanisms. Cryobiology [Internet]. 1990;**27**(3): 247-268. Available from: http://www.ncbi.nlm.nih.gov/pubmed/2199153

[6] Fahy GM. Cryoprotectant toxicity neutralization. Cryobiology [Internet]. 2010;**60**(3 Suppl):S45-S53. DOI: 10.1016/j.cryobiol.2009.05.005

[7] Nakamura Y, Obata R, Okuyama N, Aono N, Hashimoto T, Kyono K. Residual ethylene glycol and dimethyl sulphoxide concentration in human ovarian tissue during warming/thawing steps following cryopreservation. Reproductive BioMedicine Online [Internet]. 2017;**35**(3):311-313. DOI: 10.1016/j.rbmo.2017.05.016

[8] Davis JM, Rowley SD, Braine HG, Piantadosi S, Santos GW. Clinical toxicity of cryopreserved bone marrow graft infusion. Blood [Internet]. 1990; **75**(3):781-786. Available from: http://www.ncbi.nlm.nih.gov/pubmed/2297578

[9] Iwatani M, Ikegami K, Kremenska Y, Hattori N, Tanaka S, Yagi S, et al. Dimethyl sulfoxide has an impact on epigenetic profile in mouse embryoid body. Stem Cells [Internet]. 2006; **24**(11):2549-2556. DOI: 10.1634/stemcells.2005-0427

[10] Thirumala S, Goebel WS, Woods EJ. Clinical grade adult stem cell banking. Organogenesis [Internet]. Jul 2009;**5**(3): 143-154. Available from: http://www.tandfonline.com/doi/abs/10.4161/org.5.3.9811

[11] Bégin D, Lavoué J, Gérin M. La substitution des solvants par le diméthylsulfoxyde. Free Radical Biology & Medicine. 2002;**52**:80

[12] Ali J, Shelton JN. Vitrification of preimplantation stages of mouse embryos. Journal of Reproduction and Fertility. 1993;**98**:459-465

[13] Windrum P, Morris TCM. Severe neurotoxicity because of dimethyl sulphoxide following peripheral blood stem cell transplantation. Bone Marrow Transplantation [Internet]. 2003;**31**(4): 315-315. DOI: 10.1038/sj.bmt.1703848

[14] Moore K, Bonilla AQ. Cryopreservation of mammalian embryos: The state of the art. Annual Review of Biomedical Sciences. 2006;**8**:19-32

[15] Juang J-K, Liu H-J. The effect of DMSO on natural DNA conformation in enhancing transcription. Biochemical and Biophysical Research Communications [Internet]. 1987;**146** (3):1458-1464. Available from: https://linkinghub.elsevier.com/retrieve/pii/0006291X87908138

[16] Rayos AA, Takahashi Y, Hishinuma M, Kanagawa H. Quick

freezing of unfertilized mouse oocytes using ethylene glycol with sucrose or trehalose. Journal of Reproduction and Fertility [Internet]. 1994;**100**(1): 123-129. Available from: http://eutils.ncbi.nlm.nih.gov/entrez/eutils/elink.fcgi?dbfrom=pubmed&id=8182579&retmode=ref&cmd=prlinks%5Cnpapers2://publication/uuid/FA464C5D-8EA3-451B-BBB7-D41585FF6058

[17] Kuleshova LL, Shaw JM, Trounson AO. Studies on replacing most of the penetrating cryoprotectant by polymers for embryo cryopreservation. Cryobiology. 2002;**43**(1):21-31

[18] Kuleshova LL, Shaw JM. A strategy for rapid cooling of mouse embryos within a double straw to eliminate the risk of contamination during storage in liquid nitrogen. Human Reproduction. 2000;**15**(12):2604-2609

[19] Liebermann J, Dietl J, Vanderzwalmen P, Tucker MJ. Recent developments in human oocyte, embryo and blastocyst vitrification: Where are we now? Reproductive BioMedicine Online [Internet]. 2003;7(6):623-633. DOI: 10.1016/S1472-6483(10)62084-6

[20] Argyle CE, Harper JC, Davies MC. Oocyte cryopreservation: Where are we now? Human Reproduction Update [Internet]. 2016;**22**(4):440-449. DOI: 10.1093/humupd/dmw007

[21] Liu J. Development of a Novel Cryoconservation Method for Mammalian Oocytes. Szent Istvan University; 2009

[22] Yoshida S. Chemical and biophysical changes in the plasma membrane during cold acclimation of mulberry bark cells (Morus bombycis Koidz. cv Goroji). Plant Physiology [Internet]. 1984;**76**(1): 257-265. DOI: 10.1104/pp.76.1.257

[23] Thomashow MF. PLANT COLD ACCLIMATION: Freezing tolerance genes and regulatory mechanisms. Annual Review of Plant Physiology and Plant Molecular Biology [Internet]. 1999;**50**(1):571-599. DOI: 10.1146/annurev.arplant.50.1.571

[24] Nakamura K, Hatakeyama T, Hatakeyama H. Studies on bound water of cellulose by differential scanning calorimetry. Textile Research Journal. 1981;**51**(9):607-613

[25] Hatakeyema T, Yamauchi A, Hatakeyema H. Studies on bound water in poly(vinyl alcohol). Hydrogel by DSC and FT-NMR. European Polymer Journal. 1984;**20**(1):61-64

[26] Ratto JA, Hatakeyama T, Blumstein RB. Differential scanning calorimetry investigation of phase transitions in water/chitosan systems. Polymer. 1995;**36**(15):2915-2919

[27] Hatakeyama T, Hatakeyama H. Heat capacity and nuclear magnetic relaxation times of non-freezing water restrained by polysaccharides, revisited. Journal of Biomaterials Science, Polymer Edition [Internet]. 2017;**28** (10-12):1215-1230. DOI: 10.1080/09205063.2017.1291551

[28] Hatakeyama H, Onishi T, Endo T, Hatakeyama T. Gelation of chemically cross-linked methylcellulose studied by DSC and AFM. Carbohydrate Polymers. 2007;**69**(4):792-798

[29] Grossutti M, Dutcher JR. Correlation between chain architecture and hydration water structure in polysaccharides. Biomacromolecules. 2016;**17**(3):1198-1204

[30] Branca C, Magazù S, Maisano G, Migliardo F, Migliardo P, Romeo G. α,α-trehalose/water solutions. 5. Hydration and viscosity in dilute and semidilute disaccharide solutions. The Journal of Physical Chemistry B. 2001; **105**(41):10140-10145

[31] Kasaai MR. Calculation of Mark–Houwink–Sakurada (MHS) equation

viscometric constants for chitosan in any solvent–temperature system using experimental reported viscometric constants data. Carbohydrate Polymers [Internet]. 2007;**68**(3):477-488. Available from: http://linkinghub.elsevier.com/retrieve/pii/S0144861706005728

[32] Branca C, Magazù S, Maisano G, Migliardo P. Anomalous cryoprotective effectiveness of trehalose: Raman scattering evidences. The Journal of Chemical Physics. 1999;**111**(1):281-287

[33] Donnamaria MC, Howard E, Grigera JR. Interaction of water with a,a-trehalose in solution: Molecular dynamics simulation approach. Journal of the Chemical Society, Faraday Transactions. 1994;**90**(18):2731-2735

[34] Pagnotta SE, McLain SE, Soper AK, Bruni F, Ricci MA. Water and trehalose: How much do they interact with each other? The Journal of Physical Chemistry B. 2010;**114**(14):4904-4908

[35] Lerbret A, Bordat P, Affouard F, Descamps M, Migliardo F. How homogeneous are the trehalose, maltose, and sucrose water solutions? An insight from molecular dynamics simulations. The Journal of Physical Chemistry B [Internet]. 2005;**109**(21):11046-11057. DOI: 10.1021/jp0468657

[36] Hatakeyama T, Iijima M, Hatakeyama H. Role of bound water on structural change of water insoluble polysaccharides. Food Hydrocolloids [Internet]. 2016;**53**:62-68. DOI: 10.1016/j.foodhyd.2014.12.033

[37] Quan GB, Han Y, Liu MX, Fang L, Du W, Ren SP, et al. Addition of oligosaccharide decreases the freezing lesions on human red blood cell membrane in the presence of dextran and glucose. Cryobiology [Internet]. 2011;**62**(2):135-144. DOI: 10.1016/j.cryobiol.2011.01.015

[38] Watanabe K, Yaguchi T, Yang D, Kanno T, Nagai K, Yamamoto S, et al. Beneficial effect of intracellular free high-mannose oligosaccharides on cryopreservation of mammalian cells and proteins. Cryobiology. 2006;**53**(3):330-335

[39] Liu J, Tanrikut C, Wright DL, Lee GY, Toner M, Biggers JD, et al. Cryopreservation of human spermatozoa with minimal non-permeable cryoprotectant. Cryobiology [Internet]. 2016;**73**(2):162-167. DOI: 10.1016/j.cryobiol.2016.08.004

[40] Iqbal S, Naz S, Ahmed H, Andrabi SMH. Cryoprotectant effect of trehalose in extender on post-thaw quality and in vivo fertility of water buffalo (*Bubalus bubalis*) bull spermatozoa. Andrologia. 2018;**50**(1):1-6

[41] Zhu Z, Fan X, Pan Y, Lu Y, Zeng W. Trehalose improves rabbit sperm quality during cryopreservation. Cryobiology [Internet]. 2017;**75**:45-51. DOI: 10.1016/j.cryobiol.2017.02.006

[42] Pérez-Marín CC, Requena FD, Arando A, Ortiz-Villalón S, Requena F, Agüera EI. Effect of trehalose- and sucrose-based extenders on equine sperm quality after vitrification: Preliminary results. Cryobiology [Internet]. 2018;**80**:62-69. DOI: 10.1016/j.cryobiol.2017.12.002

[43] Huang H, Zhao G, Zhang Y, Xu J, Toth TL, He X. Predehydration and ice seeding in the presence of trehalose enable cell cryopreservation. ACS Biomaterials Science & Engineering. 2017;**3**(8):1758-1768

[44] Somjit K, Ruttanapornwareesakul Y, Hara K, Nozaki Y. The cryoprotectant effect of shrimp chitin and shrimp chitin hydrolysate on denaturation and unfrozen water of lizardfish surimi during frozen storage. Food Research International. 2005;**38**(4):345-355

[45] Shu GW, Hu M, Qin T, Chen H, Ma Q. Effect of fructo-oligosaccharide, isomalto-oligosaccharide, inulin and xylo-oligosaccharide on survival of *B. Bifidum* during freeze-drying. Advanced Materials Research [Internet]. 2011;**382**: 454-457. Available from: http://www.scientific.net/AMR.382.454

[46] Miyamoto Y, Suzuki S, Nomura K, Enosawa S. Improvement of hepatocyte viability after cryopreservation by supplementation of long-chain oligosaccharide in the freezing medium in rats and humans. Cell Transplantation [Internet]. 2006; **15**(10):911-919. DOI: 10.3727/000000006783981404

[47] Stushnoff C, Seufferheld MJ, Creegan T. Oligosaccharides as endogenous cryoprotectants in woody plants. In: Plant Cold Hardiness [Internet]. Boston, MA: Springer US; 1997. pp. 301-309. Available from: http://link.springer.com/10.1007/978-1-4899-0277-1_27

[48] Auh JH, Lee HG, Kim JW, Kim JC, Yoon HS, Park KH. Highly concentrated branched oligosaccharides as cryoprotectant for surimi. Journal of Food Science [Internet]. 1999;**64**(3): 418-422. DOI: 10.1111/j.1365-2621.1999.tb15055.x

[49] Aisen EG, Medina VH, Venturino A. Cryopreservation and post-thawed fertility of ram semen frozen in different trehalose concentrations. Theriogenology. 2002; **57**(7):1801-1808

[50] Aisen E, Quintana M, Medina V, Morello H, Venturino A. Ultramicroscopic and biochemical changes in ram spermatozoa cryopreserved with trehalose-based hypertonic extenders. Cryobiology. 2005;**50**(3):239-249

[51] Aboagla EM-E. Trehalose-enhanced fluidity of the goat sperm membrane and its protection during freezing. Biology of Reproduction [Internet]. 2003;**69**(4):1245-1250. DOI: 10.1095/biolreprod.103.017889

[52] Kasai M, Komi JH, Takakamo A, Tsudera H, Sakurai T, Machida T. A simple method for mouse embryo cryopreservation in a low toxicity vitrification solution, without appreciable loss of viability. Journal of Reproduction and Fertility [Internet]. 1990;**89**(1):91-97. Available from: http://www.ncbi.nlm.nih.gov/pubmed/2374136

[53] Rall WF, Wood MJ. High in vitro and in vivo survival of day 3 mouse embryos vitrified or frozen in a non-toxic solution of glycerol and albumin. Journal of Reproduction and Fertility. 1994;**101**:681-688

[54] Shaw JM, Kuleshova LL, MacFarlane DR, Trounson AO. Vitrification properties of solutions of ethylene glycol in saline containing PVP, Ficoll, or dextran. Cryobiology. 1997;**35**(3):219-229

[55] Wang Y, Okitsu O, Zhao XM, Sun Y, Di W, Chian RC. The effect of minimal concentration of ethylene glycol (EG) combined with polyvinylpyrrolidone (PVP) on mouse oocyte survival and subsequent embryonic development following vitrification. Journal of Assisted Reproduction and Genetics. 2014;**31**(1):55-63

[56] Wallach EE, Friedler S, Giudice LC, Lamb EJ. Cryopreservation of embryos and ova. Fertility and Sterility [Internet]. 1988;**49**(5):743-764. Available from: http://linkinghub.elsevier.com/retrieve/pii/S0015028216598793

[57] Balaban B, Urman B, Ata B, Isiklar A, Larman MG, Hamilton R, et al. A randomized controlled study of human day 3 embryo cryopreservation by slow freezing or vitrification: Vitrification is associated with higher survival, metabolism and blastocyst

formation. Human Reproduction. 2008; **23**(9):1976-1982

[58] Gardner DK, Lane M, Stevens J, Schoolcraft WB. Changing the start temperature and cooling rate in a slow-freezing protocol increases human blastocyst viability. Fertility and Sterility. 2003;**79**(2):407-410

[59] Rall WF, Wood MJ, Kirby C, Whittingham DG. Development of mouse embryos cryopreserved by vitrification. Journal of Reproduction and Fertility [Internet]. 1987;**80**(2): 499-504 Available from: http://www.ncbi.nlm.nih.gov/pubmed/3656282

[60] Khurana NK, Niemann H. Effects of cryopreservation on glucose metabolism and survival of bovine morulae and blastocysts derived in vitro or in vivo. Theriogenology [Internet]. 2000;**54**(2): 313-326. DOI: 10.1016/S0093-691X(00) 00351-4

[61] Barbas JP, Mascarenhas RD. Cryopreservation of domestic animal sperm cells. Cell and Tissue Banking. 2009;**10**:49-62

[62] Gil HW, Lee TH, Park I. Effects of cryoprotectants and diluents on the cryopreservation of spermatozoa from far eastern catfish, Silurus asotus. Development & Reproduction [Internet]. 2017;**21**(1):71-91. Available from: http://www.ksdb.org/archive/view_article?pid=dr-21-1-79

[63] Rall WF, Fahy GM. Cryopreservation of mouse embryos at −196 degrees C by vitrification. Nature. 1985;**313**:573-575

[64] Palasz AT, Mapletoft RJ. Cryopreservation of mammalian embryos and oocytes: Recent advances. Biotechnology Advances. 1996;**14**(2): 127-149

[65] Ali J, Shelton JN. Design of vitrification solutions for the cryopreservation of embryos. Journal of Reproduction and Fertility. 1993;**99**(2): 471-477

[66] Kasai M. Cryopreservation of mammalian embryos. Molecular Biotechnology. 1997;7:173-179

[67] Gómez MC, Pope E, Harris R, Mikota S, Dresser BL. Development of in vitro matured, in vitro fertilized domestic cat embryos following cryopreservation, culture and transfer. Theriogenology. 2003;**60**(2):239-251

[68] Eroglu A, Lawitts JA, Toner M, Toth TL. Quantitative microinjection of trehalose into mouse oocytes and zygotes, and its effect on development. Cryobiology. 2003;**46**(2):121-134

[69] Kasai M, Mukaida T. Cryopreservation of animal and human embryos by vitrification. Reproductive BioMedicine Online [Internet]. 2004: 164-170. DOI: 10.1016/S1472-6483(10) 62125-6. Reproductive Healthcare Ltd, Duck End Farm, Dry Drayton, Cambridge CB23 8DB, UK

[70] Eroglu A, Bailey SE, Toner M, Toth TL. Successful cryopreservation of mouse oocytes by using low concentrations of trehalose and dimethylsulfoxide1. Biology of Reproduction [Internet]. 2009;**80**(1): 70-78. DOI: 10.1095/biolreprod.108.070383

[71] Krag KT, Koehler I-M, Wright RW. Trehalose—A nonpermeable cryoprotectant for direct freezing of early stage murine embryos. Theriogenology [Internet]. 1985;**23**(1): 200. Available from: http://www.sciencedirect.com/science/article/pii/0011224085901348

[72] Storey BT, Noiles EE, Thompson KA. Comparison of glycerol, other polyols, trehalose, and raffinose to provide a defined cryoprotectant medium for mouse sperm

cryopreservation. Cryobiology [Internet]. 1998;**37**(1):46-58. Available from: http://www.ncbi.nlm.nih.gov/pubmed/9698429

[73] Saha S, Otoi T, Takagi M, Boediono A, Sumantri C, Suzuki T. Normal calves obtained after direct transfer of vitrified bovine embryos using ethylene glycol, trehalose, and polyvinylpyrrolidone. Cryobiology. 1996;**33**(3):291-299

[74] Wu CF, Tsung HC, Zhang WJ, Wang Y, Lu JH, Tang ZY, et al. Improved cryopreservation of human embryonic stem cells with trehalose. Reproductive BioMedicine Online [Internet]. 2005;**11**(6):733-739. DOI: 10.1016/S1472-6483(10)61692-6

[75] McGinnis LK, Zhu L, Lawitts JA, Bhowmick S, Toner M, Biggers JD. Mouse sperm desiccated and stored in trehalose medium without freezing. Biology of Reproduction [Internet]. 2005;**73**(4):627-633. DOI: 10.1095/biolreprod.105.042291

[76] Eroglu A, Toner M, Toth TL. Beneficial effect of microinjected trehalose on the cryosurvival of human oocytes. Fertility and Sterility. 2002; **77**(1):152-158

[77] Buchanan SS, Gross SA, Acker JP, Toner M, Carpenter JF, Pyatt DW. Cryopreservation of stem cells using trehalose: Evaluation of the method using a human hematopoietic cell line. Stem Cells and Development [Internet]. 2004;**13**(3):295-305. DOI: 10.1089/154732804323099226

[78] Honadel TE, Killian GJ. Cryopreservation of murine embryos with trehalose and glycerol. Cryobiology. 1988;**25**(4):331-337

[79] Judycka S, Cejko BI, Dryl K, Dobosz S, Grudniewska J, Kowalski RK. The effect of supplementation of a trehalose-based extender with KCl on rainbow trout (*Oncorhynchus mykiss*) sperm freezability and post-thaw motility. Aquaculture [Internet]. 2016; **465**:303-310. DOI: 10.1016/j.aquaculture.2016.09.029

[80] De Castro MPV, Cortell C, Vicente JS. Dextran vitrification media prevents mucin coat and zona pellucida damage in rabbit embryo. Theriogenology [Internet]. 2010;**74**(9): 1623-1628. DOI: 10.1016/j.theriogenology.2010.06.034

[81] Dumoulin JCM, Bergers-Janssen JM, Pieters MHEC, Enginsu ME, Geraedts JPM, Evers JLH. The protective effects of polymers in the cryopreservation of human and mouse zonae pellucidae and embryos. Fertility and Sterility [Internet]. 1994;**62**(4): 793-798. DOI: 10.1016/S0015-0282(16)57006-X

[82] Sieme H, Oldenhof H, Wolkers W. Sperm membrane behaviour during cooling and cryopreservation. Reproduction in Domestic Animals [Internet]. 2015;**50**:20-26. DOI: 10.1111/rda.12594

[83] Meryman HT. Review of biological freezing. In: Mery HT, editor. Cryobiology. New-York: Academic Press; 1966. pp. 1-114

[84] Ashwood-Smith MJ, Warby C. Studies on the molecular weight and cryoprotective properties of polyvinylpyrrolidone and dextran with bacteria and erythrocytes. Cryobiology. 1971;**8**(5):453-464

[85] Bailey TL, Stubbs C, Murray K, Tomás RMF, Otten L, Gibson MI. Synthetically scalable poly(ampholyte) which dramatically enhances cellular cryopreservation. Biomacromolecules [Internet]. 2019;**20**(8):3104-3114. DOI: 10.1021/acs.biomac.9b00681

[86] Prickett RC, Elliott JAW, McGann LE. Application of the osmotic

virial equation in cryobiology. Cryobiology [Internet]. 2010 Feb;**60**(1): 30-42. DOI: 10.1016/j.cryobiol.2009.07.011

[87] Fonseca F, Meneghel J, Cenard S, Passot S, Morris GJ. Determination of Intracellular vitrification temperatures for unicellular micro organisms under conditions relevant for cryopreservation. PLoS One [Internet]. 2016;**11**(4):1-19. 10.1371/journal.pone.0152939

[88] Browne RK, Clulow J, Mahony M. The effect of saccharides on the post-thaw recovery of cane toad (*Bufo marinus*) spermatozoa. Cryo Letters [Internet]. 2002;**23**(2):121-128. Available from: http://www.ncbi.nlm.nih.gov/pubmed/12050780

[89] Huang H, Choi JK, Rao W, Zhao S, Agarwal P, Zhao G, et al. Alginate hydrogel microencapsulation inhibits devitrification and enables large-volume low-CPA cell vitrification. Advanced Functional Materials. 2015;**25**(44): 6839-6850

[90] Meneghel J, Passot S, Cenard S, Réfrégiers M, Jamme F, Fonseca F. Subcellular membrane fluidity of Lactobacillus delbrueckii subsp. bulgaricus under cold and osmotic stress. Applied Microbiology and Biotechnology [Internet]. 2017;**101**(18):6907-6917. DOI: 10.1007/s00253-017-8444-9

[91] Kent B, Hunt T, Darwish TA, Hauß T, Garvey CJ, Bryant G. Localization of trehalose in partially hydrated DOPC bilayers: Insights into cryoprotective mechanisms. Journal of The Royal Society Interface [Internet]. 2014;**11** (95):20140069. DOI: 10.1098/rsif.2014.0069

[92] Deller RC, Vatish M, Mitchell DA, Gibson MI. Synthetic polymers enable non-vitreous cellular cryopreservation by reducing ice crystal growth during thawing. Nature Communications [Internet]. 2014;**5**(1):3244. Available from: http://www.nature.com/articles/ncomms4244

[93] Morris GJ, Acton E, Avery S. A novel approach to sperm cryopreservation. Human Reproduction [Internet]. 1999;**14**(4):1013-1021. Available from: http://www.ncbi.nlm.nih.gov/pubmed/10221235

[94] Fonseca F, Passot S, Lieben P, Marin M. Collapse temperature of bacterial suspensions: The effect of cell type and concentration. Cryo-Letters. 2004;**25**(6):425-434

[95] Mika JT, Van Den Bogaart G, Veenhoff L, Krasnikov V, Poolman B. Molecular sieving properties of the cytoplasm of *Escherichia coli* and consequences of osmotic stress. Molecular Microbiology. 2010;**77**(1):200-207

[96] Zhou EH, Trepat X, Park CY, Lenormand G, Oliver MN, Mijailovich SM, et al. Universal behavior of the osmotically compressed cell and its analogy to the colloidal glass transition. Proceedings of the National Academy of Sciences [Internet]. 2009; **106**(26):10632-10637. DOI: 10.1073/pnas.0901462106

[97] Mourão MA, Hakim JB, Schnell S. Connecting the dots: The effects of macromolecular crowding on cell physiology. Biophysical Journal. 2014; **107**(12):2761-2766

[98] Wolfe J, Bryant G. Cellular cryobiology: Thermodynamic and mechanical effects. International Journal of Refrigeration. 2001;**24**:438-450

[99] Meryman HT. The exceeding of a minimum tolerable cell volume in hypertonic suspension as a cause of freezing injury. In: Wolstenholme GEW, O'Connor M, editors. The Frozen Cell [Internet]. Longmon Gr.; 1970. pp. 51-67. Available from: http://doi.wiley.com/10.1002/9780470719732.ch4

[100] Mazur P, Pinn IL, Kleinhans FW. Intracellular ice formation in mouse oocytes subjected to interrupted rapid cooling. Cryobiology [Internet]. 2007; **55**(2):158-166. Available from: http://linkinghub.elsevier.com/retrieve/pii/S0011224007001022

[101] Seki S, Mazur P. Effect of warming rate on the survival of vitrified mouse oocytes and on the recrystallization of intracellular Ice1. Biology of Reproduction [Internet]. 2008;**79**(4):727-737. DOI: 10.1095/biolreprod.108.069401

[102] Generalic E. Acetyl glucosamine [Internet]. Glossary, Croatian-English Chemistry Dictionary. 2017. Available from: https://glossary.periodni.com/dictionary.php?en=aceyl+glucosamine

[103] Xia W, Liu P, Zhang J, Chen J. Biological activities of chitosan and chitooligosaccharides. Food Hydrocolloids [Internet]. 2011;**25**(2): 170-179. DOI: 10.1016/j.foodhyd.2010.03.003

[104] Kim SK, Rajapakse N. Enzymatic production and biological activities of chitosan oligosaccharides (COS): A review. Carbohydrate Polymers. 2005; **62**(4):357-368

[105] Aam BB, Heggset EB, Norberg AL, Sørlie M, Vårum KM, Eijsink VGH. Production of chitooligosaccharides and their potential applications in medicine. Marine Drugs. 2010;**8**(5):1482-1517

[106] Escobedo-Lozano AY, Domard A, Velázquez CA, Goycoolea FM, Argüelles-Monal WM. Physical properties and antibacterial activity of chitosan/acemannan mixed systems. Carbohydrate Polymers [Internet]. 2015;**115**:707-714. DOI: 10.1016/j.carbpol.2014.07.064

[107] Peter MG. Applications and environmental aspects of chitin and chitosan. Journal of Macromolecular Science, Part A [Internet]. 1995;**32**(4): 629-640. DOI: 10.1080/10601329508010276

[108] Adekunle KF. Bio-based polymers for technical applications: A review—Part 2. Open Journal of Polymer Chemistry [Internet]. 2014;**04**(04):95-101. DOI: 10.4236/ojpchem.2014.44011

[109] Teixeira M, Buff S, Desnos H, Loiseau C, Bruyère P, Joly T, et al. Ice nucleating agents allow embryo freezing without manual seeding. Theriogenology [Internet]. 2017; **104**(104):173-178 Available from: http://linkinghub.elsevier.com/retrieve/pii/S0093691X17303977

[110] Moussa A, Crépet A, Ladavière C, Trombotto S. Reducing-end "clickable" functionalizations of chitosan oligomers for the synthesis of chitosan-based diblock copolymers. Carbohydrate Polymers [Internet]. 2019;**219**:387-394. Available from: https://linkinghub.elsevier.com/retrieve/pii/S0144861719304783

[111] Becerra J, Sudre G, Royaud I, Montserret R, Verrier B, Rochas C, et al. Tuning the hydrophilic/hydrophobic balance to control the structure of chitosan films and their protein release behavior. AAPS PharmSciTech [Internet]. 2017;**18**(4):1070-1083. DOI: 10.1208/s12249-016-0678-9

[112] Dhawade PP, Jagtap RN. Characterization of the glass transition temperature of chitosan and its oligomers by temperature modulated differential scanning calorimetry. Advances in Applied Science Research [Internet]. 2012;**3**(3):1372-1382. Available from: www.pelagiaresearchlibrary.com

[113] Payet L. Viscoelasticite et structure de gels à base de chitosane - relations avec les propriétés diffusionnelles de macromolécules dans ces biogels. Solutions. Université Paris 7 - Denis Diderot; 2005

[114] Barros SC, da Silva AA, Costa DB, Costa CM, Lanceros-Méndez S, Maciavello MNT, et al. Thermal–mechanical behaviour of chitosan–cellulose derivative thermoreversible hydrogel films. Cellulose. 2015;**22**(3): 1911-1929

[115] Julkapli NM, Akil HM. Influence of a plasticizer on the mechanical properties of kenaf-filled chitosan bio-composites. Polymer-Plastics Technology and Engineering. 2010;**49**(9):944-951

[116] El-Hefian EA, Elgannoudi ES, Mainal A, Yahaya AH. Characterization of chitosan in acetic acid: Rheological and thermal studies. Turkish Journal of Chemistry. 2010;**34**(1):47-56

[117] Desnos H, Baudot A, Teixeira M, Louis G, Commin L, Buff S, et al. Ice induction in DSC experiments with Snomax®. Thermochimica Acta [Internet]. 2018;**667**:193-206 Available from: https://linkinghub.elsevier.com/retrieve/pii/S0040603118305938

[118] Boutron P. More accurate determination of the quantity of ice crystallized at low cooling rates in the glycerol and 1,2-propanediol aqueous solutions: Comparison with equilibrium. Cryobiology. 1984;**21**(2):183-191

[119] Desnos H. Amélioration des procédures de cryoconservation de type congélation-lente par simulation et caractérisation des effets de composés chitooligosaccharides. University Lyon 1; 2019

[120] Angell CA, Shuppert J, Tucker JC. Anomalous properties of supercooled water. Heat capacity, expansivity, and proton magnetic resonance chemical shift from 0 to −38°C. The Journal of Physical Chemistry [Internet]. 1973; **77**(26):3092-3099. DOI: 10.1021/j100644a014

[121] Weast RC. CRC Handbook of Chemistry and Physics. 60th ed. In: Weast RC, editor. CRC Press; 1979

Chapter 2

Cryoprotection of Platelets by Grafted Polymers

Mark D. Scott, Nobu Nakane and Elisabeth Maurer-Spurej

Abstract

Unlike red blood cells (RBC) which are stored at 4°C, platelets are stored at 22–24°C (room temperature) due to biophysical and biochemical changes induced by cold temperatures aggregately known as the 'cold storage lesion' (CSL). However, 22°C storage greatly increases the risk of microbial growth, thus limiting the safe storage of platelets to only 5–7 days (versus 42 days for RBC). Consequent to the short shelf life of platelets, blood services face chronic shortages of these life-saving cells. To overcome both the risk of microbial contamination and the constrained supplies of platelets, renewed research into attenuating the CSL and/or determining where cold stored platelets are clinically suitable are ongoing. In this chapter, we show that the covalent grafting of methoxypolyethylene glycol (mPEG), a biocompatible polymer, to the membrane of platelets attenuates the CSL. Moreover, the grafted mPEG serves as a potent cryoprotectant allowing platelets to be stored at 4°C, or frozen at −20°C, while retaining normal platelet counts and biologic function. The successful development of platelet PEGylation may provide a means by which the cold storage of platelets can be achieved with a minimal loss of platelet quality while improving both platelet microbial safety and inventory.

Keywords: cryopreservation, cryoprotection, platelets, blood banking, cold storage, PEGylation, immunocamouflage, methoxypoly(ethylene glycol), polymer, aggregation

1. Introduction

Platelet adhesion and aggregation at the site of vascular injury are key events required for normal vascular homeostasis and wound repair. [1–4] Platelets are produced from megakaryocytes in the bone marrow and, while lacking a nucleus, contain a number of specialized granules such as alpha-granules and dense granules. Normal, resting platelets have a discoid morphology which changes upon activation to 'spiny spheres' arising from the formation of pseudopodia. This shape change coincides with the rearrangement of the actin cytoskeleton. Upon activation, platelets adhere to the subendothelium at sites of vascular injury, aggregate and initiate coagulation to stop bleeding (*i.e.* haemostasis).

Consequent to this essential role, platelet transfusions have evolved as a crucial therapeutic tool in the treatment of a large number of diverse clinical conditions including acute bleeding, surgery, treatment of a variety of cancers, patients with platelet abnormalities and autoimmune diseases such as Idiopathic Thrombocytopenic Purpura (ITP) [5]. To meet the increasing clinical needs, blood systems within developed countries produce in excess of 5,000,000 transfusion doses annually [6]. However, demand for platelets continues to increase annually while the rate of blood/platelet donations are actually declining leading to an inventory that is chronically constrained [7].

The constraint of platelet inventory is in large part due to an inability to safely store platelet products for greater than 5–7 days. Historically, platelets, like red blood cells (RBC), were stored at 4°C and successfully used clinically. However, multiple studies from the late 1960s to the early 1970s demonstrated that 4°C (*i.e.* cold) storage of platelets resulted in significantly reduced *in vivo* survival times compared to platelets stored at 22°C (warm storage) or endogenously produced platelets (2–4 vs. 7–9 vs. 10–12 days, respectively) [8–17]. The observed loss of *in vivo* viability and *in vitro* morphology and function was termed the platelet cold storage lesion (CSL) and resulted in the change in standard blood banking practice to storing platelets at 22–24°C by the early 1970s.

The CSL is multi-dimensional and is best characterized as the sum of all the deleterious changes in platelet morphology, biochemistry and function that arise from the time the blood is withdrawn to the time the cold-stored platelets are transfused. The CSL is characterized, in part, by loss of discoid shape (*i.e.* abnormal morphological forms), decreased mean platelet volume, increased size

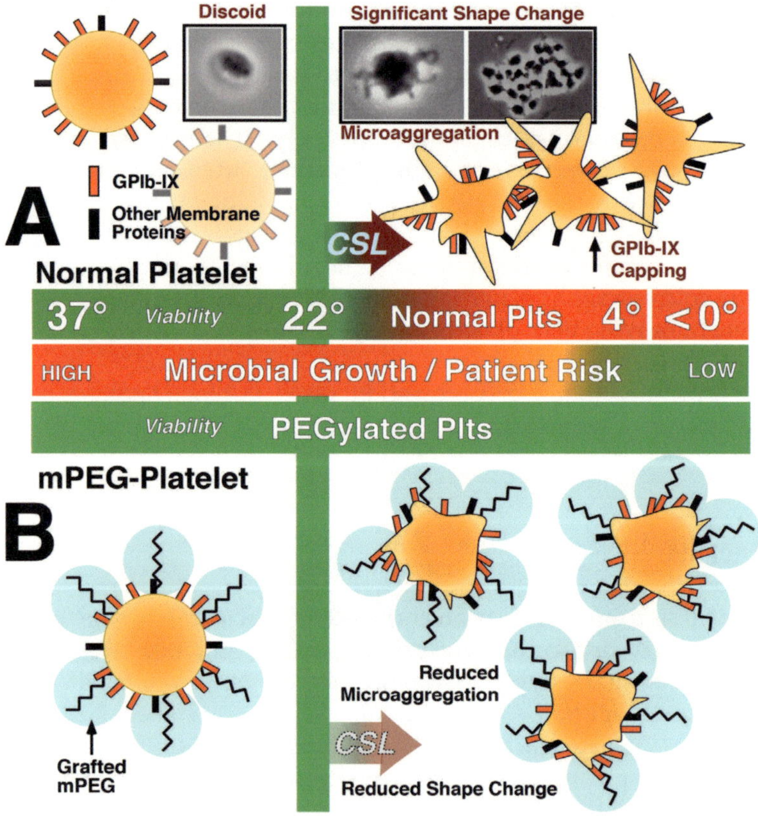

Figure 1.
*Schematic view of the prevention of the platelet 'cold storage lesion' (CSL) by membrane PEGylation. Panel A: The normal discoid shape of platelets is lost as platelets are cooled below ~18°C. As a consequence of cooling, pseudopodia formation occurs leading to microaggregation of platelets (photo insert). Additionally, membrane proteins such as GP1b-IX aggregate on the surface. These changes lead to both mechanical and immunological clearance from the circulation. Panel B: PEGylation of platelets reduces both the shape change (*e.g. *fewer pseudopodia) and prevents microaggregation of the cold stored platelets. Consequent to the attenuation of the CSL lesions by the grafted mPEG, platelets can be stored for extended periods (> 7 days) at 4 or −80°C thereby improving platelet inventory and supply management while reducing the platelet discard rates. As noted, cold storage also significantly reduces the risk of microbial growth thus, potentially, improving transfusion safety.*

heterogeneity, pseudopodia formation, increased release of platelet α-granules and cytosolic proteins, altered surface protein expression (*e.g.*, glycoproteins such as GP1b-IX), increased procoagulant activity, aggregate formation, and reduced platelet counts—all of which are also characteristic of platelet activation (**Figure 1A**) [8–16, 18, 19]. In contrast, warm storage of platelets maintained platelet morphology, activation potential and greatly improved *in vivo* circulation times [8–16].

However, the warm storage of platelets was not without risk as it was demonstrated that warm storage significantly increased the risk for bacterial growth should bacteria be introduced to the platelet unit during collection [20–24]. Indeed, numerous North American screening studies have indicated that approximately 1/3500 platelet units (primarily platelet rich plasma; PRP) are bacterially contaminated posing a potential hazard to already at-risk patients [23, 25, 26]. Consequent to this risk, multiple blood systems have implemented costly universal bacteriologic screening of donor platelets. Hence, development of new technologies to improve both platelet safety and inventories will be crucial in meeting the ever increasing demand for platelet products.

2. What was 'OLD' is 'NEW' again

Consequent to the clinical demand and supply chain issues, several studies over the last several years have re-explored the potential use of 'cold-stored' platelets. Initial excitement regarding cold-stored platelets arose in 2003, Hoffmeister et al. investigated the mechanism(s) underlying the CSL and experimentally demonstrated that the shape change alone induced by cold storage itself did not result in poor platelet survival in a murine model [18, 19]. Instead, Hoffmeister et al. hypothesized that poor platelet survival resulted from an irreversible membrane clustering of alpha subunits of glycoprotein Ib (GPIbα). Their studies reported that exposed, terminal, beta-linked N-acetylglucosamine (βGlcNAc) residues on clustered GPIbα were recognized by the lectin domain of type 3 complement receptors (CR3; $\alpha_M\beta 2$; CD11b/CD18) on liver and splenic macrophages. This immunorecognition resulted in the rapid clearance of cold stored donor platelets via phagocytosis. Hoffmeister also demonstrate that phagocytosis of briefly chilled murine platelets could be inhibited and *in vivo* survival prolonged by enzymatically galactosylating the terminal βGlcNAc residues on GPIbα. These findings led them to propose that enzymatically masking the exposed ßGlcNAc residues on the N-glycans of the clustered GPIbα molecules by galactosylation would allow for the cold storage of human platelets without adversely affecting platelet function. However, enthusiasm for glycosylated platelets subsided when subsequent studies by Wandall et al. demonstrated that galactosylation alone did NOT protect murine or human platelets from prolonged cold storage (*e.g.* >48 hours) [27].

More recently, the 'old' (1960) has become 'new' (2019) as transfusion scientists have begun to reexamine the clinical utility of platelets stored at 4°C. Indeed, the original 1960's/70's studies that initially discovered the platelet CSL, also reported that 'cold-stored' platelets were still effective *in vivo* in preventing acute blood loss. Hence, current clinical studies are investigating the use of 4°C stored platelets for at least some transfusion demands [28–45]. In general, recent studies suggest that, while these 'old technology' cold-stored platelets could be of benefit for acute haemostatic transfusion needs, cold stored platelets still exhibit morphological changes and poor *in vivo* survival making them unlikely candidates for chronic replacement therapy in patients with already accelerated platelet clearance or as therapeutics for patients with cancer or who have undergone bone marrow transplantation.

3. Hypothesis: attenuating the CSL via membrane-grafted mPEG

Consequent to our earlier work on polymer grafting to intact cells (*e.g.* RBC, lymphocytes), we hypothesized that the polymer induced immunocamouflage of platelet membranes with methoxypoly(ethylene glycol) [mPEG] could prevent or circumvent the immune recognition of cold stored platelets [46–49]. This hypothesis was supported by our previous studies on RBC and leukocytes (White blood cells; WBC) that demonstrated that the grafted polymer prevented cell:cell interactions (*e.g.* RBC Rouleaux formation; Phagocytosis of opsonized RBC; and Lymphocyte:APC) and membrane protein clustering (RBC CR1 aggregation) of the mPEG-modified cells while maintaining normal cellular function [50–71]. Hence, it was hypothesized that mPEG-grafting to platelets would prevent platelet aggregation and the clustering of GPIbα, and phagocytic recognition of the transfused mPEG-platelets (**Figure 1B**). Moreover, because soluble mPEG is a known cryoprotectant, we hypothesized that the grafted polymer might also attenuate other 'cold-induced' mechanical lesions of the CSL induced by cold temperatures and even freezing of donor platelets [72–79]. Indeed, the ability to freeze and recover donor platelets would both greatly increase platelet inventory and potentially expand the use of platelet therapy to geographic locations where platelet therapy is not commonly practiced.

4. Polymer engineering of platelets

All human experiments were done in accordance with the approval of the University of British Columbia Clinical Research Ethics Board and the Canadian Blood Services Ethics Review Board in accordance with the Code of Ethics of the World Medical Association (Declaration of Helsinki). Following informed consent, fresh platelet rich plasma (PRP) and, in some cases, buffy coat platelets (the standard of care in Canada and Western Europe) were obtained from volunteer donors or the Canadian Blood Services Network Centre for Applied Development (NetCAD) Laboratory (Vancouver, BC). PRP samples are similar to the platelet preparations used clinically and have an advantage of a lower level of manipulation (*e.g.* centrifugation) than buffy coat or apheresis platelets. Care was taken to assure adequate representation of males and females and no individuals were excluded on the basis of age (within acceptable age range of donation of 17–71) or race.

Based on our previous studies, platelets were PEGylated using a semi-automated PEGylation device to maintain a constant platelet:polymer ratio **(Figure 2)** in a micromixing chamber to assure uniform polymer grafting [50, 52, 56, 71]. Platelets were modified with monofunctional (*i.e.* one binding site per chain) mPEG-succinimidyl propionate (SPA-mPEG). Previous studies (not shown) within our laboratories determined that the optimal molecular weights for the PEGylation of platelets were 2–5 kDa which were used in these studies. For comparison of the effects of different linker chemistry, some studies simultaneously examined mPEG-benzotriazolyl carbonate (BTC-mPEG). Both SPA- and BTC-mPEG (Laysan Bio, Inc., Arab, AL, USA) react with protein lysine residues and covalently attach via formation of a stable amide bond (**Figure 2C**). To demonstrate the efficacy of the polymer-mediated grafting, the mPEG-mediated immunocamouflage of platelet CD9 was assessed by flow cytometry. CD9 is a constitutive tetraspan membrane glycoprotein present on resting platelets that modulates cell adhesion and migration.

Control and PEGylated mini-units (approximately 50 ml/unit; ~500 × 10^9/L) of platelets were stored at 4 or 22°C with agitation per Canadian Blood Services standard operating procedures. Storage at −80°C was done separately in the sample

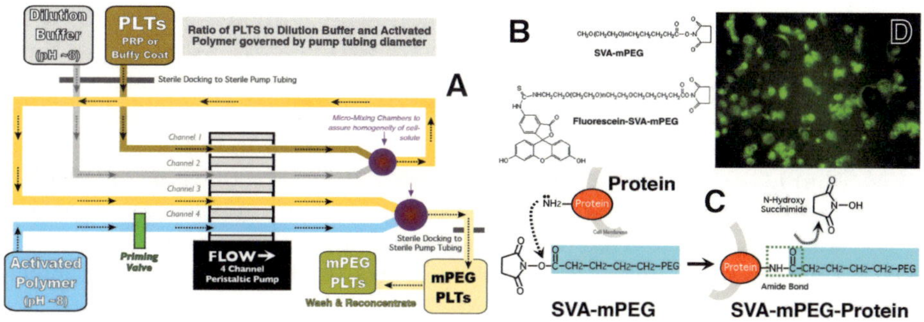

Figure 2.
Production of mPEG-platelets and the SVAmPEG reaction scheme and grafting efficacy. Panel A: semi-automated PEGylation device allowing for control of the mPEG:platelet ratio for uniform grafting levels [71]. Panels B and C: the structure (B) and reaction scheme (C) for activated SVA-mPEG. The SVA-linker chemistry forms a stable amide bond with platelet membrane proteins. Panel D: Fluorescein-conjugated SVA-mPEG demonstrated that platelets were uniformly modified using our semi-automated methodology.

mini-unit blood banking bags. Storage was done for up to 12 days under the prescribed conditions (note: normal storage at 22°C is only allowed for 7 days). Storage bags were sampled aseptically in biosafety cabinets; washing and lysis procedures were performed as described previously [80–82]. Platelet counts were determined using an Advia 120 Hematology Analyzer (Bayer Inc., Toronto, Canada).

5. Effect of 4°C storage on mPEG-platelets

The covalent grafting of mPEG to PRP platelets resulted in the efficient immunocamouflage of CD9 (**Figure 3A**). As demonstrated in **Figure 3A**, virtually 100% of control platelets were $CD9^{+}$, while the mPEG grafting to the platelets exhibited dose effect on the immunocamouflage of CD9. More importantly, the grafted polymer significantly decreased the aggregation of human platelets at 4°C. As microscopically demonstrated in **Figure 3B**, temperature exerted a significant effect on the morphology and microaggregation of control PRP preparations. As anticipated, minimal differences were observed in the control platelets at 37° (*i.e. in vivo* conditions) versus 22°C (normal *in vitro* storage temperature). Importantly, PEGylation with either SCmPEG5000 or BTCmPEG5000 yielded platelets with comparable morphology to the control cells at 37 and 22°C. However, upon thermal transition from 22 to 4°C, control platelets were observed to form significant microaggregates characteristic of the CSL. In stark contrast, neither the SCmPEG5000 nor BTCmPEG5000 modified platelets exhibited any significant microaggregation consequent to the mPEG-mediated inhibition of cell:cell interaction [66, 68]. Morphological analysis of the SCmPEG5000 and BTCmPEG5000 modified platelets suggested that SCmPEG5000 better prevented cold induced shape change relative to the BTC polymer resulting in the SC-linker chemistry being further explored.

The mPEG-mediated inhibition of cold-induced platelet aggregation was also not a short term effect. As demonstrated in **Figure 4**, unmodified control platelets demonstrated significant shape change, microaggregation, and a dramatic (~30%) decrease in platelet count. In contrast, minimal microaggregation was noted in the PEGylated samples following 12 days storage at 4°C. PEGylated platelets also retained a more discoid shape (though some pseudopod formation was noted). Due to the inhibition of microaggregation and inhibition of activation induced shape change, the mPEG-grafted platelets also resulted in a significantly improved platelet count.

Importantly, PEGylated platelets were functionally normal as evidenced by their *in vitro* aggregation response to thrombin. As shown in **Figure 5A**, phase contrast microscopy of washed control and PEGylated platelets resuspended in normal plasma both maintained a smooth, resting morphology. However, in response to 2 IU/mL thrombin activation (**Figure 5B**), both control (a) and PEGylated (b; 10 mM SCmPEG-5000) platelets fully aggregated at 37°C (1000 rpm stir speed) in an aggregometer (Chronolog, Havertown, PA, USA). This normal aggregation of PEGylated platelets occurred despite the significant immunocamouflage (see below) of the platelet membrane surface. Moreover, as shown in **Figure 5C**, phase contrast microscopy demonstrated that control and PEGylated platelets form microscopically very similar thrombin-induced clots. Interestingly, if PEGylated platelets were suspended in PEGylated plasma essentially no *in vitro* aggregation was observed, likely due to the PEGylation of plasma proteins involved in

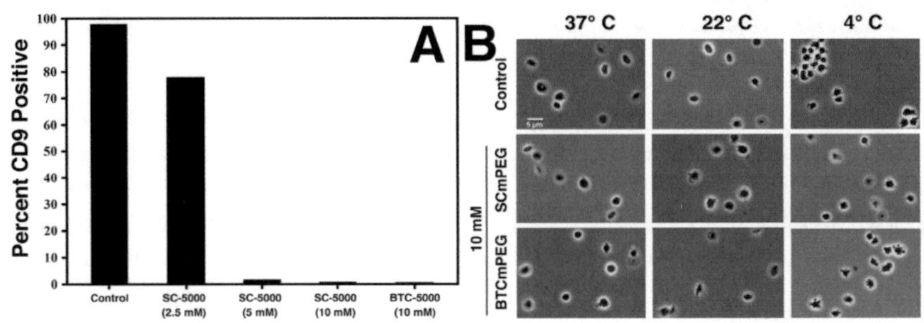

Figure 3.
Immunocamouflage of platelets by grafted mPEG. Panel A: CD9 is effectively camouflaged by grafted 5 kDa polymers of SC- and BTC-activated mPEG. The efficacy of immunocamouflage was a function of mPEG grafting concentration. Panel B: While both SC- and BTC-mPEG demonstrated similar efficiency in camouflaging CD9, photomicrographs showed that the SCmPEG better preserved platelet morphology at 4°C. SCmPEG was consequently used for all further studies.

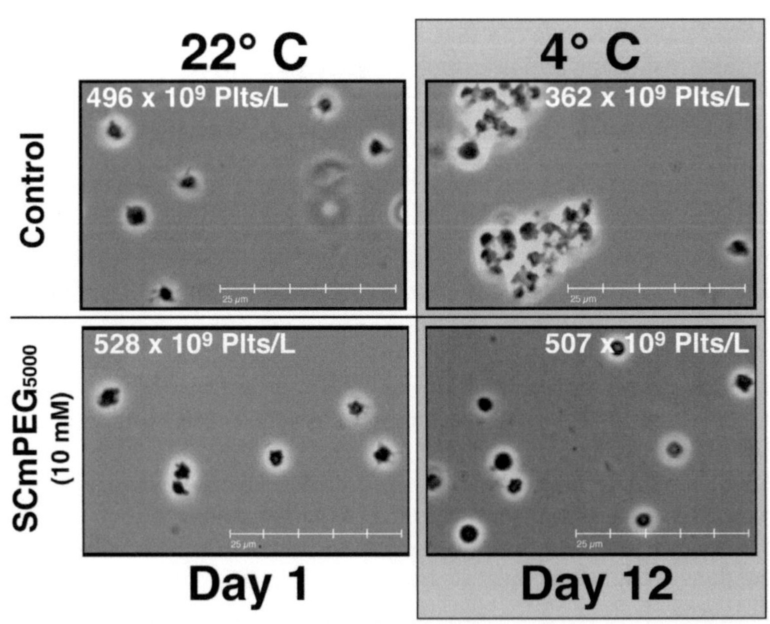

Figure 4.
PEGylation inhibited 4°C cold-induced platelets aggregation and shape change. The improved viability of the cells is accompanied by maintenance of the pre-storage platelet count (day 0).

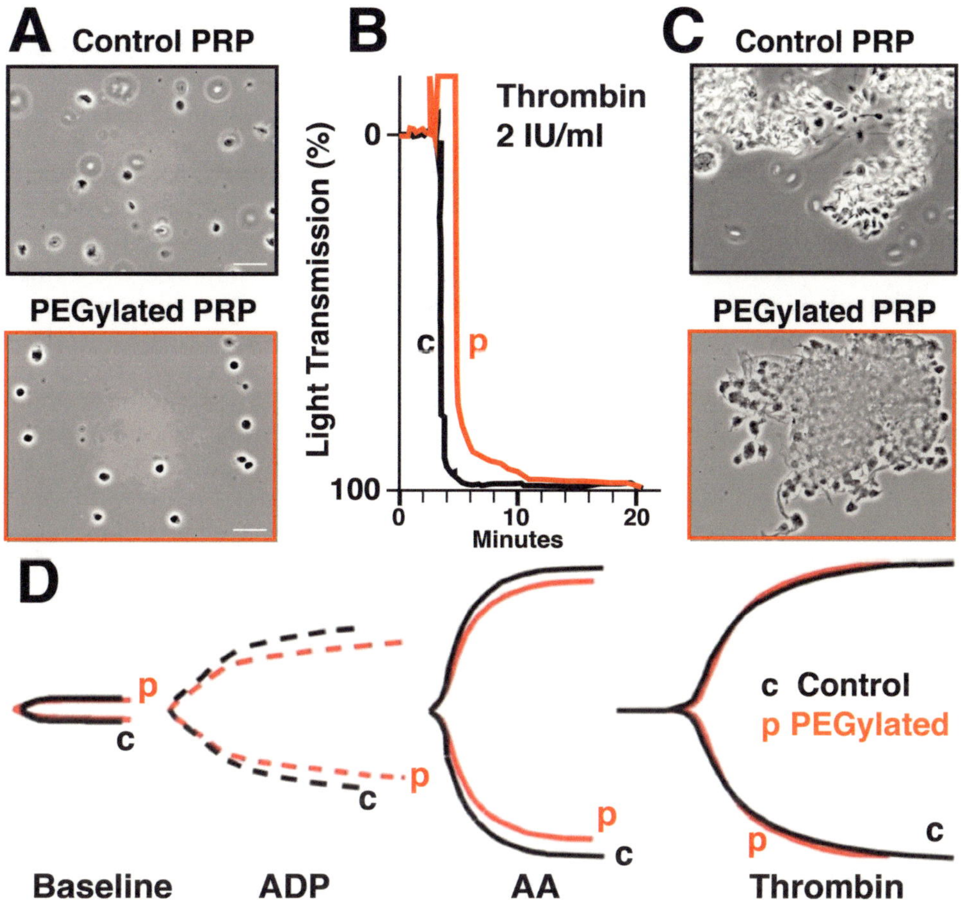

Figure 5.
PEGylated platelets are functional and aggregate in vitro *in response to agonists. Panel A: Phase contrast microscopy of control and PEGylated PRP platelets in plasma demonstrate that both populations maintain a smooth, resting morphology. Panel B: Aggregometer analysis of control (c) and PEGylated (p) platelets demonstrate normal responses to 2 IU/mL thrombin (37°C; 1000 rpm). Panel C: Control and PEGylated PRP platelets form microscopically very similar thrombin-induced clots. Panel D: PEGylation did not affect platelet thromboelastography (TEG) as denoted by the virtually identical TEG tracings. These results demonstrated that PEGylated platelets participated normally in the in vitro clotting assay. Platelet mapping with TEG determines total platelet function. The two symmetric arms show the same results. Shown are representative responses of control (c; black lines) and PEGylated (red lines; SCmPEG5000) platelets at rest (baseline) and in response to ADP, AA, and thrombin activation.*

clot formation. However, these data indicated that PEGylated platelets would be functional when transfused into an actively bleeding patient regardless of whether they were stored in the presence of PEGylated plasma, normal plasma or a platelet storage solution as their functionality was restored in the presence of normal plasma. Thromboelastographic (Haemonetics, Braintree, MA) analysis of control and PEGylated platelets further demonstrated the normality of polymer modified platelets in response to multiple platelet agonists. As noted in **Figure 5D**, control and PEGylated platelets demonstrated virtually identical results when exposed to adenosine diphosphate (ADP), arachidonic acid (AA) or thrombin activation indicating that clot formation should not be adversely affected by PEGylation—especially since in most circumstances PEGylated donor platelets will represent ~50% or, most typically, much less of the platelets in a clot.

Clinically, visual inspection for 'swirl' may be the only pre-transfusion 'quality' test of the platelet unit—though even this is rarely done. The swirl test is a noninvasive method that literally works by swirling the bag and looking for light diffraction (*i.e.* refractiveness) [83–86]. Due to the discoid shape of resting (unactivated) platelets, light is diffracted creating a cloud- or swirl-like appearance of the bag. Platelet activation causes a disc to sphere morphology change where upon orientation dependent changes in light diffraction are no longer observed. A dull platelet bag is deemed 'activated' while a refractive bag is 'resting'. Unsurprisingly, despite the low cost (*i.e.* free) of the swirl test, it actually tells very little as to the quality of the platelet unit. Over the last few years, a new technology has been developed to quantitatively measure platelet quality using dynamic light scattering. The ThromboLUX (LightIntegra Technology, Vancouver, BC) quantitatively measures platelet morphology (shape change) and temperature response and provides a quantitative replacement to the qualitative and subjective 'platelet swirl' [62, 83–85, 87–92]. Mechanistically the ThromboLUX utilizes dynamic light scattering to examine a small volume (~30 μl) of platelets to quantitatively assess platelet size and morphological changes arising from temperature cycling (37 to 20 to 37°C). The ThomboLUX generates a Dynamic Light Scattering (DLS) value that correlates with platelet activation status. Moreover, the ThromboLUX is capable of quantitating the number of platelet-derived microparticles and evidence of microbial contamination. The ThromboLUX technology has been clinically validated and provides a correlation between the DLS score and a patients corrected count increment at 24 hours (CCI24) post transfusion [84]. Shown in **Figure 6A** is the ThromboLUX

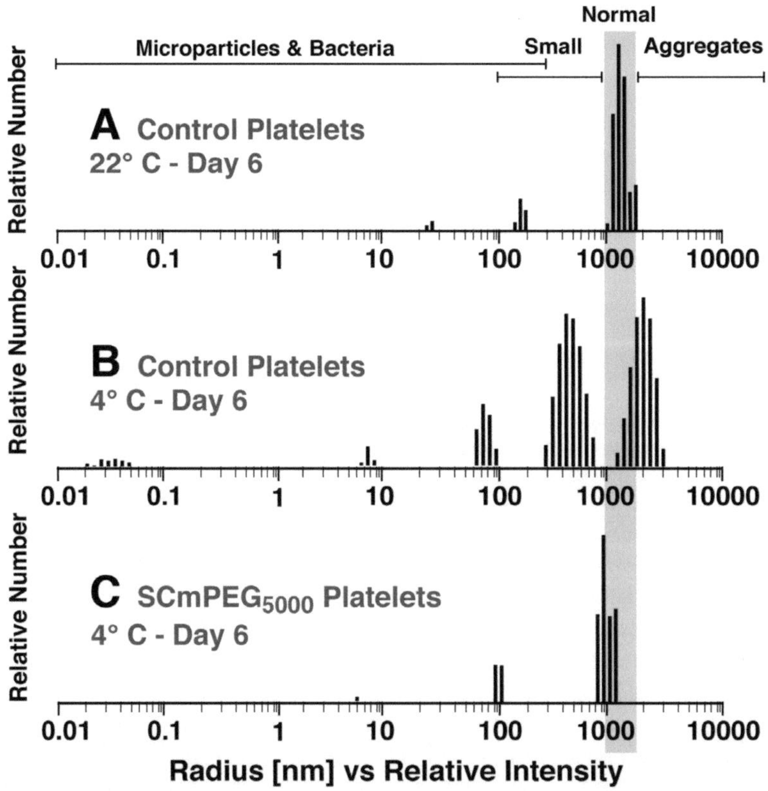

Figure 6.
ThromboLUX dynamic light scattering analysis of platelets following 6 days storage. Panel A: Control platelets stored at 22°C. panel B: Control platelets stored at 4°C. panel C: ScmPEG5000 (10 mM) platelets stored at 4°C. As shown, mPEG grafting resulted in significant cryoprotection. The Y-axis is particle count while the x-axis reflects the hydrodynamic radius distribution. All samples were prepared from the same donor platelet unit.

profile of control platelets stored 6 days at 22°C. However, cold storage of the same platelet preparation resulted in a dramatic shift of the platelet peak (**Figure 6B**) with the appearance of microparticles, smaller platelets and platelet aggregates. In contrast, cold stored (6 days) PEGylated platelets (10 mM, 5 kDa) yielded a DLS profile similar to the 22°C stored platelets with no evidence of aggregate formation and minimal microparticle formation (**Figure 6C**).

6. Effect of −80°C storage on mPEG-platelets

To further assess the cryoprotective effects of the grafted mPEG polymer, freezing studies were conducted on the control and SCmPEG platelets. While previous work on PEG as a cryoprotectant utilized a soluble form, work with PEG and other cryoprotectants (DMSO and Trehalose) demonstrated that the primary site of protection was at the level of the cell membrane [72–79]. As shown in **Figure 7**, following 12 days storage at −80°C, covalently bound SCmPEG provided significant cryoprotection as reflected by both platelet morphology and improved cell counts. This finding is in stark contrast to control platelets which exhibited significant fragmentation and dramatically reduced cell counts post storage and thawing.

The covalent grafted mPEG exerted additional benefits post thaw. As shown in **Figure 8**, SCmPEG-grafted platelets exhibited improved morphology, and less fragmentation, immediately post-thaw when compared to control cells. Indeed, the grafted polymer provided comparable (or better) cryoprotection than DMSO. Moreover, following washing and re-concentration of the freeze-thaw

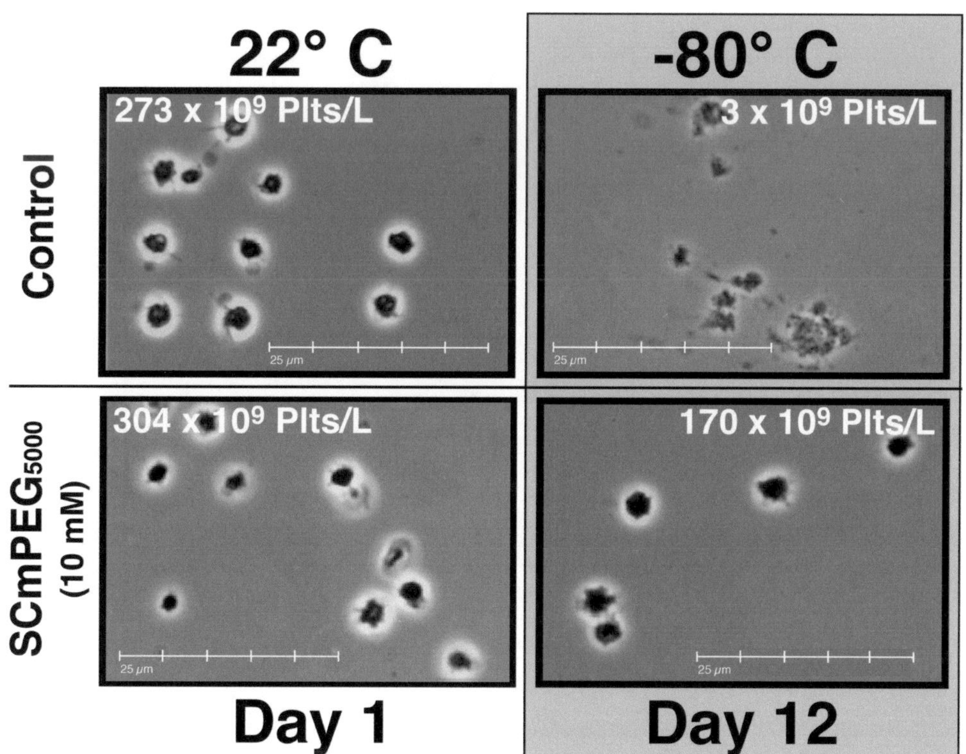

Figure 7.
PEGylation inhibited −80°C cold-induced platelets aggregation and shape change. The improved viability of the cells is accompanied by significantly improved maintenance of the pre-storage platelet count (day 0).

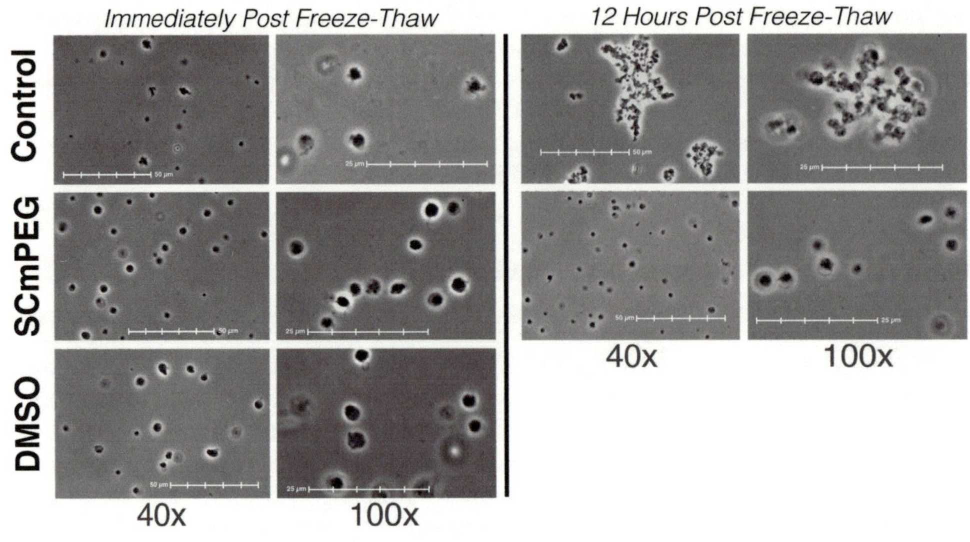

Figure 8.
PEGylation gives rise to significant cryoprotection from freeze-thaw injury. As shown, after 12 days of storage at −80°C, unmodified control platelets exhibited significant platelet destruction, loss of morphology, and a significant ($p < 0.001$) decrease in platelet count. In contrast, subsequent to freeze-thaw, PEGylated platelets demonstrated normal discoid morphology and few detectable microaggregates and were very comparable to platelets cryopreserved with DMSO, the standard cryoprotectant for frozen RBC and platelets. Moreover, unlike control platelets, the PEGylated sample did not form microaggregates when incubated at 22°C.

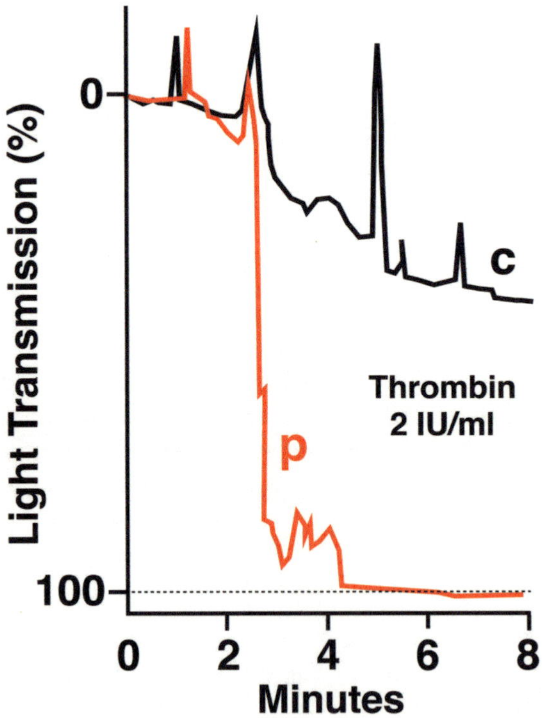

Figure 9.
Post storage at −80°C, the aggregation of control and SCmPEG platelets was assessed in response to thrombin (2 IU/mL). As shown, thawed control platelets exhibited limited aggregation (~40% light transmittance) after 8 minutes. In contrast, the PEGylated platelets demonstrated robust aggregation (100% light transmittance at ~4 minutes). Shown are representative responses of control (black lines) and PEGylated (red lines; SCmPEG5000) platelets.

control and SCmPEG platelets, control platelets demonstrated significant aggregation when stored at 22°C overnight (12 hours). In contrast, the SCmPEG-platelets demonstrated no aggregation over the same 12 hour time frame.

While the maintenance of morphology and platelet numbers post −80 storage was promising, the key question was whether these platelets were functional. To assess platelet function, thrombin (2 IU/ml) induced aggregation was assessed. As shown in **Figure 9**, thawed control platelets exhibited a poor response to thrombin (see **Figure 5B** for a normal response) as demonstrated by very limited aggregation. Moreover, the aggregation of the control platelets was very slow as seen by the slope of the aggregation curve. In contrast, the thawed SCmPEG platelets demonstrated significant, and rapid, thrombin mediated aggregation. Indeed, near maximal aggregation was achieved within approximately 3 minutes and very closely resembled the thrombin activation curves of fresh control and SCmPEG PRP (see **Figure 5B**).

7. Discussion

Platelet transfusions are a critical component in the treatment of both traumatic acute injury and a number of chronic diseases. However, unlike RBC which are stored at 4°C, platelets are stored at 22–24°C (room temperature) due to the induction of the CSL at temperatures below ~18°C. While the CSL encompasses a multitude of biophysical and biochemical changes, perhaps the most apparent effect is the production of platelet aggregates. To prevent the CSL, blood services worldwide have successfully stored platelets at 22°C. However, warm storage has its own risks as it greatly increases the risk for microbial growth limiting the safe storage of platelets to only 5–7 days (versus 42 days for RBC) and the outdating of a significant number of donor units. Consequent to the short shelf life of platelets, blood services face chronic shortages of these life-saving cells. To overcome both the risk of microbial contamination and the constrained supplies of platelets, renewed research into attenuating the CSL and/or determining where cold stored platelets are clinically suitable are ongoing.

To circumvent the microbial risk, and improve platelet inventory, our research has examined the potential use of cold stored, mPEG-grafted, platelets. As demonstrated by our *in vitro* experimental findings, the covalent grafting of mPEG to donor platelets significantly reduced the severity of the 4°C CSL while maintaining normal haemostatic function. This was evidenced by the maintenance of 'normal' platelet morphology, the lack of microaggregation, and normal platelet activation by thrombin and normal aggregation as determined by thromboelastography and aggregation studies. Moreover, due to the cryoprotective effects of the grafted mPEG, polymer-modified platelets could be stored at −80°C and thawed and still retained normal morphology and haemostatic function. In contrast, the freezing of normal platelets in the absence of a cryoprotectant (*e.g.* DMSO) resulted in significant cellular damage resulting in vastly reduced recovery, loss of haemostatic function and subsequent microaggregation when incubated at room temperature for 12 hours.

Interestingly, while not a focus of this chapter, polymer size (*e.g.* 2, 5 or 20 kDa) was an important factor when considering the cryoprotection of platelets while maintaining the unique functions of the platelet. In contrast to RBC and WBC in which longer chain polymers (*e.g.* 20 kDa) were optimal, short chain polymers (*e.g.* 2–5 kDa) were optimal for platelet cryopreservation. However, the disparity between RBC and WBC relative to platelets is not as unanticipated as it appears.

The goal of RBC and WBC PEGylation is to induce immunocamouflage (*i.e.* prevent immune recognition) and, especially for WBC, prevent cell:cell communication (*e.g.* allorecognition). Clearly for platelets, cell:cell interaction is crucial for their haemostatic function. Hence, the primary goal of platelet PEGylation is to prevent thermal injury to the membrane while maintaining normal platelet function. While all sizes of the grafted polymer could prevent cryogenic injury, long chain polymers inhibited platelet activation and their subsequent aggregation. In contrast, the short 2–5 kDa polymers provided adequate cryoprotective effects while allowing for normal agonist-driven activation.

8. Conclusions

PEGylation of donor platelets with short chain (2–5 kDa) mPEG effectively prevent the overt morphological changes arising from the CSL. Moreover, the polymer-grafted platelets retained their normal haemostatic function following both cold storage (4°C) and freezing (−80°C) as evidenced by thromboelastography and aggregation studies. Importantly, cold storage of platelets would improve transfusion safety as it would diminish the risk of microbial growth in a blood product destined for use in at risk patients. Also of potential clinical and economic importance was the finding that mPEG-grafted platelets withstood freezing in the absence of other cryoprotectants such as DMSO. The use of frozen platelets, requiring no DMSO removal step, could expand the availability of platelet transfusions to geographic regions in which they are not currently available or where donor recruitment or production facilities do not exist. The successful implementation of this technology for the cold storage of platelets would be of significant benefit to transfusion recipients by increasing the availability of platelets for transfusion.

Acknowledgements

This work was supported by grants from the Canadian Blood Services (MDS; EMS and MDS) and Health Canada (MDS). The views expressed herein do not necessarily represent the view of the federal government of Canada. We thank the Canada Foundation for Innovation and the Michael Smith Foundation for Health Research for infrastructure funding at the University of British Columbia Centre for Blood Research. The funders had no role in study design, data collection and analysis, decision to publish, or preparation of the manuscript.

Conflict of interest

The Canadian Blood Services (Ottawa, ON, Canada) has patents relating to the cold storage of platelets [48, 49]. MDS, NN and EMS are inventors cited on said patents.

Author details

Mark D. Scott[1,2,3*], Nobu Nakane[1] and Elisabeth Maurer-Spurej[1,2,3]

1 Canadian Blood Services, Ottawa, ON, Canada

2 The Centre for Blood Research, University of British Columbia, Vancouver, BC, Canada

3 Department of Pathology and Laboratory Medicine, University of British Columbia, Vancouver, BC, Canada

*Address all correspondence to: mdscott@mail.ubc.ca

References

[1] Slichter SJ. Platelet transfusion therapy. Hematology/Oncology Clinics of North America. 1990;**4**:291-311. DOI: 10.1016/S0889-8588(18)30517-3

[2] Murphy S, Varma M. Selecting platelets for transfusion of the alloimmunized patient: A review. Immunohematology. 1998;**14**:117-123

[3] Hartwig JH. The platelet: Form and function. Seminars in Hematology. 2006;**43**:S94-S100. DOI: 10.1053/j.seminhematol.2005.11.004

[4] Wohner N. Role of cellular elements in thrombus formation and dissolution. Cardiovascular & Hematological Agents in Medicinal Chemistry. 2008;**6**:224-228. DOI: 10.2174/187152508784871972

[5] Cobain TJ, Vamvakas EC, Wells A, Titlestad K. A survey of the demographics of blood use. Transfusion Medicine. 2007;**17**:1-15. DOI: 10.1111/j.1365-3148.2006.00709.x

[6] Stroncek DF, Rebulla P. Platelet transfusions. Lancet. 2007;**370**:427-438. DOI: 10.1016/S0140-6736(07)61198-2

[7] Goldman M, Steele WR, Di Angelantonio E, van den Hurk K, Vassallo RR, Germain M, et al. Biomedical EFSTCBESTI. Comparison of donor and general population demographics over time: A BEST collaborative group study. Transfusion. 2017;**57**:2469-2476. DOI: 10.1111/trf.14307

[8] Murphy S, Gardner FH. Effect of storage temperature on maintenance of platelet viability--deleterious effect of refrigerated storage. The New England Journal of Medicine. 1969;**280**:1094-1098. DOI: 10.1056/NEJM196905152802004

[9] Murphy S, Gardner FH. The effect of temperature on platelet viability. Vox Sanguinis. 1969;**17**:22

[10] Murphy S, Sayar SN, Gardner FH. Storage of platelet concentrates at 22 degrees C. Blood. 1970;**35**:549-557

[11] Murphy S, Gardner FH. Maintenance of platelet viability and functional integrity during storage. Vox Sanguinis. 1971;**20**:427-428. DOI: 10.1111/j.1423-0410.1971.tb01814.x

[12] Murphy S, Gardner FH. Platelet storage at 22 degrees C; metabolic, morphologic, and functional studies. The Journal of Clinical Investigation. 1971;**50**:370-377. DOI: 10.1172/JCI106504

[13] Becker GA, Tuccelli M, Kunicki T, Chalos MK, Aster RH. Studies of platelet concentrates stored at 22 C and 4 C. Transfusion. 1973;**13**:61-68. DOI: 10.1111/j.1537-2995.1973.tb05442.x

[14] Slichter SJ, Harker LA. Preparation and storage of platelet concentrates. Transfusion. 1976;**16**:8-12

[15] Holme S, Vaidja K, Murphy S. Platelet storage at 22 degrees C: Effect of type of agitation on morphology, viability, and function in vitro. Blood. 1978;**52**:425-435

[16] Slichter SJ. Preservation of platelet viability and function during storage of concentrates. Progress in Clinical and Biological Research. 1978;**28**:83-100

[17] Winokur R, Hartwig JH. Mechanism of shape change in chilled human platelets. Blood. 1995;**85**:1796-1804

[18] Hoffmeister KM, Josefsson EC, Isaac NA, Clausen H, Hartwig JH, Stossel TP. Glycosylation restores survival of chilled blood platelets. Science. 2003;**301**:1531-1534. DOI: 10.1126/science.1085322

[19] Hoffmeister KM, Felbinger TW, Falet H, Denis CV, Bergmeier W,

Mayadas TN, et al. The clearance mechanism of chilled blood platelets. Cell. 2003;**112**:87-97. DOI: 10.1016/S0092-8674(02)01253-9

[20] Wagner SJ, Friedman LI, Dodd RY. Transfusion-associated bacterial sepsis. Clinical Microbiology Reviews. 1994;7:290-302. DOI: 10.1128/CMR.7.3.290

[21] Blajchman MA. Bacterial contamination of blood products and the value of pre-transfusion testing. Immunological Investigations. 1995;**24**:163-170. DOI: 10.3109/08820139509062770

[22] Dumont LJ, AuBuchon JP, Whitley P, Herschel LH, Johnson A, McNeil D, et al. Seven-day storage of single-donor platelets: Recovery and survival in an autologous transfusion study. Transfusion. 2002;**42**:847-854. DOI: 10.1046/j.1537-2995.2002.00147.x

[23] Hillyer CD, Josephson CD, Blajchman MA, Vostal JG, Epstein JS, Goodman JL. Bacterial contamination of blood components: Risks, strategies, and regulation: Joint ASH and AABB educational session in transfusion medicine. Hematology. American Society of Hematology. Education Program. 2003;**2003**:575-589. DOI: 10.1182/asheducation-2003.1.575

[24] Benjamin RJ, Wagner SJ. The residual risk of sepsis: Modeling the effect of concentration on bacterial detection in two-bottle culture systems and an estimation of false-negative culture rates. Transfusion. 2007;**47**:1381-1389. DOI: 10.1111/j.1537-2995.2007.01326.x

[25] Blajchman MA, Goldman M, Baeza F. Improving the bacteriological safety of platelet transfusions. Transfusion Medicine Reviews. 2004;**18**:11-24. DOI: 10.1016/j.tmrv.2003.10.002

[26] Blajchman MA, Beckers EA, Dickmeiss E, Lin L, Moore G, Muylle L. Bacterial detection of platelets: Current problems and possible resolutions. Transfusion Medicine Reviews. 2005;**19**:259-272. DOI: 10.1016/j.tmrv.2005.05.002

[27] Wandall HH, Hoffmeister KM, Sorensen AL, Rumjantseva V, Clausen H, Hartwig JH, et al. Galactosylation does not prevent the rapid clearance of long-term, 4 degrees C-stored platelets. Blood. 2008;**111**:3249-3256. DOI: 10.1182/blood-2007-06-097295

[28] Milford EM, Reade MC. Comprehensive review of platelet storage methods for use in the treatment of active hemorrhage. Transfusion. 2016;**56**(Suppl 2):S140-S148. DOI: 10.1111/trf.13504

[29] Johnson L, Tan S, Wood B, Davis A, Marks DC. Refrigeration and cryopreservation of platelets differentially affect platelet metabolism and function: A comparison with conventional platelet storage conditions. Transfusion. 2016;**56**:1807-1818. DOI: 10.1111/trf.13630

[30] Getz TM, Montgomery RK, Bynum JA, Aden JK, Pidcoke HF, Cap AP. Storage of platelets at 4°C in platelet additive solutions prevents aggregate formation and preserves platelet functional responses. Transfusion. 2016;**56**:1320-1328. DOI: 10.1111/trf.13511

[31] Spinella PC, Cap AP. Whole blood: Back to the future. Current Opinion in Hematology. 2016;**23**:536-542. DOI: 10.1097/MOH.0000000000000284

[32] Stubbs JR, Tran SA, Emery RL, Hammel SA, Haugen AL, Zielinski MD, et al. Cold platelets for trauma-associated bleeding: Regulatory approval, accreditation approval, and practice implementation-just the "tip of

the iceberg". Transfusion. 2017;**57**: 2836-2844. DOI: 10.1111/trf.14303

[33] Berzuini A, Spreafico M, Prati D. One size doesn't fit all: Should we reconsider the introduction of cold-stored platelets in blood bank inventories. F1000Res. 2017;**6**:95. DOI: 10.12688/f1000research.10363.1

[34] Wu X, Darlington DN, Montgomery RK, Liu B, Keesee JD, Scherer MR, et al. Platelets derived from fresh and cold-stored whole blood participate in clot formation in rats with acute traumatic coagulopathy. British Journal of Haematology. 2017;**179**:802-810. DOI: 10.1111/bjh.14999

[35] Stolla M, Fitzpatrick L, Gettinger I, Bailey SL, Pellham E, Christoffel T, et al. In vivo viability of extended 4°C-stored autologous apheresis platelets. Transfusion. 2018;**58**:2407-2413. DOI: 10.1111/trf.14833

[36] Ng MSY, Tung JP, Fraser JF. Platelet storage lesions: What more do we know now. Transfusion Medicine Reviews. 2018;**32**:144-154. DOI: 10.1016/j.tmrv.2018.04.001

[37] Humbrecht C, Kientz D, Gachet C. Platelet transfusion: Current challenges. Transfusion Clinique et Biologique. 2018;**25**:151-164. DOI: 10.1016/j.tracli.2018.06.004

[38] Waters L, Cameron M, Padula MP, Marks DC, Johnson L. Refrigeration, cryopreservation and pathogen inactivation: An updated perspective on platelet storage conditions. Vox Sanguinis. 2018;**113**:317-328. DOI: 10.1111/vox.12640

[39] Reddoch-Cardenas KM, Sharma U, Salgado CL, Montgomery RK, Cantu C, Cingoz N, et al. An in vitro pilot study of apheresis platelets collected on Trima Accel system and stored in T-PAS+ solution at refrigeration temperature (1-6°C). Transfusion. 2019;**59**:1789-1798. DOI: 10.1111/trf.15150

[40] Reddoch-Cardenas KM, Bynum JA, Meledeo MA, Nair PM, Wu X, Darlington DN, et al. Cold-stored platelets: A product with function optimized for hemorrhage control. Transfusion and Apheresis Science. 2019;**58**:16-22. DOI: 10.1016/j.transci.2018.12.012

[41] Leeper CM, Yazer MH, Cladis FP, Saladino R, Triulzi DJ, Gaines BA. Cold-stored whole blood platelet function is preserved in injured children with hemorrhagic shock. Journal of Trauma and Acute Care Surgery. 2019;**87**:49-53. DOI: 10.1097/TA.0000000000002340

[42] Braathen H, Sivertsen J, Lunde THF, Kristoffersen EK, Assmus J, Hervig TA, et al. In vitro quality and platelet function of cold and delayed cold storage of apheresis platelet concentrates in platelet additive solution for 21 days. Transfusion. 2019;**58**:2652-2661. DOI: 10.1111/trf.15356

[43] Scorer T, Williams A, Reddoch-Cardenas K, Mumford A. Manufacturing variables and hemostatic function of cold-stored platelets: A systematic review of the literature. Transfusion. 2019;**59**: 2722-2732. DOI: 10.1111/trf.15396

[44] Getz TM. Physiology of cold-stored platelets. Transfusion and Apheresis Science. 2019;**58**:12-15. DOI: 10.1016/j.transci.2018.12.011

[45] Ketter PM, Kamucheka R, Arulanandam B, Akers K, Cap AP. Platelet enhancement of bacterial growth during room temperature storage: Mitigation through refrigeration. Transfusion. 2019;**59**:1479-1489. DOI: 10.1111/trf.15255

[46] Scott MD, Eaton JW. Antigenic modulation of cells. US Patent Number: 5,908,624. Albany, NY, USA: Assignee Albany Medical College. 1999

[47] Scott MD, Eaton JW. Antigenic modulation of cells. US Patent Number: 8,007,784. Assignee Albany Medical College. 2011

[48] Scott MD, Maurer E. Cold Storage Of Modified Platelets At >0°C. US Patent Number: 7,964,339. Assignee Canadian Blood Services. 2011

[49] Maurer E, Scott MD, Kitamura N. Cold storage of pegylated platelets at about or below 0°C. US Patent Number: 8,067,151. Assignee Canadian Blood Services. 2011

[50] Scott MD, Murad KL, Koumpouras F, Talbot M, Eaton JW. Chemical camouflage of antigenic determinants: Stealth erythrocytes. Proceedings of the National Academy of Sciences of the United States of America. 1997;**94**:7566-7571. DOI: 10.1073/pnas.94.14.7566

[51] Murad KL, Gosselin EJ, Eaton JW, Scott MD. Stealth cells: Prevention of major histocompatibility complex class II-mediated T-cell activation by cell surface modification. Blood. 1999;**94**:2135-2141

[52] Murad KL, Mahany KL, Brugnara C, Kuypers FA, Eaton JW, Scott MD. Structural and functional consequences of antigenic modulation of red blood cells with methoxypoly(ethylene glycol). Blood. 1999;**93**:2121-2127

[53] Scott MD, Bradley AJ, Murad KL. Camouflaged blood cells: Low-technology bioengineering for transfusion medicine? Transfusion Medicine Reviews. 2000;**14**:53-63. DOI: 10.1016/S0887-7963(00)80115-7

[54] Chen AM, Scott MD. Current and future applications of immunological attenuation via pegylation of cells and tissue. BioDrugs. 2001;**15**:833-847. DOI: 10.2165/00063030-200115120-00005

[55] Bradley AJ, Test ST, Murad KL, Mitsuyoshi J, Scott MD. Interactions of IgM ABO antibodies and complement with methoxy-PEG-modified human RBCs. Transfusion. 2001;**41**:1225-1233. DOI: 10.1046/j.1537-2995.2001.41101225.x

[56] Bradley AJ, Murad KL, Regan KL, Scott MD. Biophysical consequences of linker chemistry and polymer size on stealth erythrocytes: Size does matter. Biochimica et Biophysica Acta. 2002;**1561**:147-158

[57] Bradley AJ, Scott MD. Separation and purification of methoxypoly(ethylene glycol) grafted red blood cells via two-phase partitioning. Journal of Chromatography. B, Analytical Technologies in the Biomedical and Life Sciences. 2004;**807**:163-168. DOI: 10.1016/j.jchromb.2004.03.054

[58] Chen AM, Scott MD. Comparative analysis of polymer and linker chemistries on the efficacy of immunocamouflage of murine leukocytes. Artificial Cells, Blood Substitutes, and Immobilization Biotechnology. 2006;**34**:305-322. DOI: 10.1080/10731190600683845

[59] Bradley AJ, Scott MD. Immune complex binding by immunocamouflaged [poly(ethylene glycol)-grafted] erythrocytes. American Journal of Hematology. 2007;**82**:970-975. DOI: 10.1002/ajh.20956

[60] Le Y, Scott MD. Immunocamouflage: The biophysical basis of immunoprotection by grafted methoxypoly(ethylene glycol) (mPEG).

Acta Biomaterialia. 2010;**6**:2631-2641. DOI: 10.1016/j.actbio.2010.01.031

[61] Wang D, Toyofuku WM, Chen AM, Scott MD. Induction of immunotolerance via mPEG grafting to allogeneic leukocytes. Biomaterials. 2011;**32**:9494-9503. DOI: 10.1016/j.biomaterials.2011.08.061

[62] Greco CA, Maurer-Spurej E, Scott MD, Kalab M, Nakane N, Ramirez-Arcos SM. PEGylation prevents bacteria-induced platelet activation and biofilm formation in platelet concentrates. Vox Sanguinis. 2011;**100**:336-339. DOI: 10.1111/j.1423-0410.2010.01419.x

[63] Wang D, Kyluik DL, Murad KL, Toyofuku WM, Scott MD. Polymer-mediated immunocamouflage of red blood cells: Effects of polymer size on antigenic and immunogenic recognition of allogeneic donor blood cells. Science China. Life Sciences. 2011;**54**:589-598. DOI: 10.1007/s11427-011-4190-x

[64] Le Y, Li L, Wang D, Scott MD. Immunocamouflage of latex surfaces by grafted methoxypoly(ethylene glycol) (mPEG): Proteomic analysis of plasma protein adsorption. Science China. Life Sciences. 2012;**55**:191-201. DOI: 10.1007/s11427-012-4290-2

[65] Wang D, Toyofuku WM, Scott MD. The potential utility of methoxypoly(ethylene glycol)-mediated prevention of rhesus blood group antigen RhD recognition in transfusion medicine. Biomaterials. 2012;**33**:3002-3012. DOI: 10.1016/j.biomaterials.2011.12.041

[66] Kyluik-Price DL, Li L, Scott MD. Comparative efficacy of blood cell immunocamouflage by membrane grafting of methoxypoly(ethylene glycol) and polyethyloxazoline. Biomaterials. 2014;**35**:412-422. DOI: 10.1016/j.biomaterials.2013.09.016

[67] Li L, Noumsi GT, Kwok YY, Moulds JM, Scott MD. Inhibition of phagocytic recognition of anti-D opsonized Rh D+ RBC by polymer-mediated immunocamouflage. American Journal of Hematology. 2015;**90**:1165-1170. DOI: 10.1002/ajh.24211

[68] Kyluik-Price DL, Scott MD. Effects of methoxypoly (ethylene glycol) mediated immunocamouflage on leukocyte surface marker detection, cell conjugation, activation and alloproliferation. Biomaterials. 2016;**74**:167-177. DOI: 10.1016/j.biomaterials.2015.09.047

[69] Le Y, Toyofuku WM, Scott MD. Immunogenicity of murine mPEG-red blood cells and the risk of anti-PEG antibodies in human blood donors. Experimental Hematology. 2017;**47**:36-47.e2. DOI: 10.1016/j.exphem.2016.11.001

[70] Kang N, Toyofuku WM, Yang X, Scott MD. Inhibition of allogeneic cytotoxic T cell (CD8(+)) proliferation via polymer-induced Treg (CD4(+)) cells. Acta Biomaterialia. 2017;**57**:146-155. DOI: 10.1016/j.actbio.2017.04.025

[71] Scott M, Toyofuku W, Yang X, Raj M, Kang N. Immunocamouflaged RBC for alloimmunized patients. In: Koopman-van Gemert A, editor. Transfusion Medicine and Scientific Developments. Croatia: INTECH; 2017. pp. 23-42. DOI: 10.5772/intechopen.68647

[72] Takahashi T, Hirsh A, Erbe E, Williams RJ. Mechanism of cryoprotection by extracellular polymeric solutes. Biophysical Journal. 1988;**54**:509-518. DOI: 10.1016/S0006-3495(88)82983-7

[73] Tormanen CD. Cryoprotection of purified rat kidney transamidinase by polyethylene glycol.

Cryobiology. 1992;**29**:511-518. DOI: 10.1016/0011-2240(92)90054-6

[74] Banker MC, Layne JRJ, Hicks GLJ, Wang TC. Freezing preservation of the mammalian cardiac explant. II. Comparing the protective effect of glycerol and polyethylene glycol. Cryobiology. 1992;**29**:87-94. DOI: 10.1016/0011-2240(92)90008-P

[75] Banker MC, Layne JRJ, Hicks GLJ, Wang T. Freezing preservation of the mammalian heart explant. III. Tissue dehydration and cryoprotection by polyethylene glycol. The Journal of Heart and Lung Transplantation. 1992;**11**:619-623

[76] Coundouris JA, Grant MH, Engeset J, Petrie JC, Hawksworth GM. Cryopreservation of human adult hepatocytes for use in drug metabolism and toxicity studies. Xenobiotica. 1993;**23**:1399-1409. DOI: 10.3109/00498259309059449

[77] Tsitsanou KE, Oikonomakos NG, Zographos SE, Skamnaki VT, Gregoriou M, Watson KA, et al. Effects of commonly used cryoprotectants on glycogen phosphorylase activity and structure. Protein Science. 1999;**8**:741-749. DOI: 10.1110/ps.8.4.741

[78] Mi Y, Wood G, Thoma L. Cryoprotection mechanisms of polyethylene glycols on lactate dehydrogenase during freeze-thawing. The AAPS Journal. 2004;**6**:e22. DOI: 10.1208/aapsj060322

[79] Chen YF, Tate MW, Gruner SM. Facilitating protein crystal cryoprotection in thick-walled plastic capillaries by high-pressure cryocooling. Journal of Applied Crystallography. 2009;**42**:525-530. DOI: 10.1107/S0021889809011315

[80] Thon JN, Schubert P, Duguay M, Serrano K, Lin S, Kast J, et al. Comprehensive proteomic analysis of protein changes during platelet storage requires complementary proteomic approaches. Transfusion. 2008;**48**:425-435. DOI: 10.1111/j.1537-2995.2007.01546.x

[81] Thon JN, Schubert P, Devine DV. Platelet storage lesion: A new understanding from a proteomic perspective. Transfusion Medicine Reviews. 2008;**22**:268-279. DOI: 10.1016/j.tmrv.2008.05.004

[82] Schubert P, Thon JN, Walsh GM, Chen CH, Moore ED, Devine DV, et al. A signaling pathway contributing to platelet storage lesion development: Targeting PI3-kinase-dependent Rap1 activation slows storage-induced platelet deterioration. Transfusion. 2009;**49**:1944-1955. DOI: 10.1111/j.1537-2995.2009.02224.x

[83] Maurer-Spurej E, Brown K, Labrie A, Marziali A, Glatter O. Portable dynamic light scattering instrument and method for the measurement of blood platelet suspensions. Physics in Medicine and Biology. 2006;**51**:3747-3758. DOI: 10.1088/0031-9155/51/15/010

[84] Maurer-Spurej E, Chipperfield K. Past and future approaches to assess the quality of platelets for transfusion. Transfusion Medicine Reviews. 2007;**21**:295-306. DOI: 10.1016/j.tmrv.2007.05.005

[85] Maurer-Spurej E, Labrie A, Pittendreigh C, Chipperfield K, Smith C, Heddle N, et al. Platelet quality measured with dynamic light scattering correlates with transfusion outcome in hematologic malignancies. Transfusion. 2009;**49**:2276-2284. DOI: 10.1111/j.1537-2995.2009.02302.x

[86] Maurer-Spurej E, Pittendreigh C, Yakimec J, De Badyn MH, Chipperfield K. Erroneous automated optical platelet counts in 1-hour post-transfusion blood samples. International Journal of Laboratory Hematology. 2010;**32**:e1-e8. DOI: 10.1111/j.1751-553X.2008.01097.x

[87] Xu Y, Nakane N, Maurer-Spurej E. Novel test for microparticles in platelet-rich plasma and platelet concentrates using dynamic light scattering. Transfusion. 2011;**51**:363-370. DOI: 10.1111/j.1537-2995.2010.02819.x

[88] Labrie A, Marshall A, Bedi H, Maurer-Spurej E. Characterization of platelet concentrates using dynamic light scattering. Transfusion Medicine and Hemotherapy. 2013;**40**:93-100. DOI: 10.1159/000350362

[89] Maurer-Spurej E, Chipperfield K. Could microparticles Be the universal quality indicator for platelet viability and function. Journal of Blood Transfusion. 2016;**2016**:6140239. DOI: 10.1155/2016/6140239

[90] Maurer-Spurej E, Larsen R, Labrie A, Heaton A, Chipperfield K. Microparticle content of platelet concentrates is predicted by donor microparticles and is altered by production methods and stress. Transfusion and Apheresis Science. 2016;**55**:35-43. DOI: 10.1016/j.transci.2016.07.010

[91] Kanzler P, Mahoney A, Leitner G, Witt V, Maurer-Spurej E. Microparticle detection to guide platelet management for the reduction of platelet refractoriness in children - a study proposal. Transfusion and Apheresis Science. 2017;**56**:39-44. DOI: 10.1016/j.transci.2016.12.016

[92] Millar D, Murphy L, Labrie A, Maurer-Spurej E. Routine screening method for microparticles in platelet transfusions. Journal of Visualized Experiments. 2018:e56893. DOI: 10.3791/56893

Chapter 3

Cryomedia Formula: Cellular Molecular Perspective

Noha A. Al-Otaibi

Abstract

The growing market of cell therapy medicinal products (CTMPs) and biopharmaceuticals demand effective cryopreservation with greater safety, of which the currently available cryoprotective agents [CPAs (e.g., dimethyl sulfoxide, glycerol, trehalose, etc.)] alone are unable to provide. This is due to the need of applying high concentration of CPAs to achieve verification that concomitant oxidative damages. Formulating cocktail of compounds with anti-freezing and antioxidants properties found to be advantageous to overcome the resultant damages. Each cocktail, however, demonstrate overlapping and/or unique protective and modulation effect patterns. The advance technology and research tools (e.g., OMICs) provide a deep insight on how the formulation of cryomedia can influence the cellular pathways and molecular interactions. In fact, this shed the light over the uniqueness of cryomedia formulation and how can they serve various application purposes.

Keywords: mammalian cells, cryoprotective agents, biological profile, toxicity, protection, formula, additive agents

1. Introduction

Cryopreservation is one of the most effective techniques that widely used for preserving living cells and organs in research and therapeutic industries [1]. The principle of cryopreservation is to protect cells from the application of super low temperature and ice crystal formation by using media that consist of antifreezing or cryoprotective (CPA) substances such as; glycerol, dimethylsulfoxide (DSMO) or trehalose. The expansion in clinical experiments for medical applications revealed the limitations of utilizing the conventional CPAs which resulting sub-optimal cell quality. This is attributed to the detrimental effects of conventional CPAs and their molecular interactions that compromise cell quality. The new research areas and advanced techniques significantly increase the demand of better cryopreservation to maintain the quality and functionality of cells and tissues.

Current trends in cryopreservation are actively focusing on identifying a safe and effective alternative CPAs to substitute or support the conventional agents. In addition, there are various cell types valuable for investigation and medical development and their different biological profile and functional mechanisms required customizing cryopreservation. However, there are limiting number of studies addressing the influence of the cryomedia formulation on the global proteomic and biomolecular profile of the cells.

2. Conventional cryomedia compositions and protection mechanisms

Cryomedia formulation usually encompass of cryoprotective agents (CPAs) and carrier media prepared at or close to cell isotonic concentrations to provide support to cells at low temperature [2]. It also may contain salts, pH buffers, osmotic agents, nutrients, antioxidants or apoptosis inhibitors [2]. There are about 56 CPAs commonly used for different cell cryopreservation [3, 4]. Glycerol and dimethylsulfoxide (DMSO) are the most common CPAs used in cryomedia formula. CPAs are classified based on the permeability through cell membrane into, permeable (pCPA), and impermeable (ipCPA).

pCPAs are generally small, non-ionic molecules that are highly soluble in water, even at low temperatures. They can pass through cellular membranes and equilibrate within the cytoplasm in exchange for intracellular water during dehydration without over dehydrating the cell [5]. They become solid at a lower temperature than water freezing point and subsequently suppress ice crystal formation [6] and mitigate cellular physical damage that could occur in cellular compartments and membranes. Moreover, pCPAs reduce salt-induced stress by dissolving solute and reducing concentrations in the remaining water fraction intracellularly until the cell is cooled to a sufficiently low temperature [6, 7]. pCPA permeability is controlled by their viscosity and the membrane properties of the cell itself [8, 9]. The latter mentioned is variable between different cell types as well as the varying ages of cells [10, 11]. Most of the pCPAs are polyols, such as glycerol and dimethylsulfoxide (DMSO), which are prominence in cryopreservation. Many successful cryopreserving protocols utilized these compounds for their high efficiency in compare to others such as methanol and ethylene glycol [6].

ipCPAs are large molecules usually comprised of long chains of polymers that are unable to permeate through cellular membranes. They are water soluble and thought to increase the osmolarity around cells, which result in cellular dehydration and reduce ice crystal formation intracellularly [12]. The combination of high concentrations of ipCPAs and fast cooling promotes vitrification and stabilizing cellular proteins and membranes [13, 14]. Their protective mechanism is based on preventing ice formation extracellularly as well as intracellular through dehydration [15]. There are several classes of ipCPAs, such as certain forms of sugars, macromolecules, and polymers. Sugars are classified based on their chemical structure into: mono-, di- and poly-saccharides (glucose, trehalose, and raffinose, respectively). A number of these sugars are permeable (e.g., glucose) and others are impermeable (e.g., trehalose). Sugars have garnered unique interest over the last decades. They have been found to protect protein activity and reduce thermal denaturing heat capacity of chemicals [16–21], which leads to protein stabilization. In particular, trehalose has been identified as a universal protein stabilizer and been involved in many freezing and desiccation studies [18–20].

Depending on the freezing mode (slow cooling or vitrification), the concentrations of CPAs are variant in the solution. For instance, slow freezing mode requires less CPA concentration than vitrification. Choosing the optimum concentration of CPA in combination with the cooling rate is crucial for successful cryopreservation [20]. For instance, when preserving human ovarian tissue following slow freezing protocol, DMSO is used with initial concentration of 7.5% and gradually increased to 12.5%. Whereas preserving the same tissue using the vitrification protocol, 20% DMSO is needed [21]. Different tissues and cells, however, demonstrate different responses to the cryopreservation approaches and CPAs. Therefore, the selection of the appropriate protocol and CPA is subject to the cryopreservation empirical success of the desired cells or tissues.

3. Quality assessment methods of cryopreserved cells and CPA protection action

An accurate assessment of cryopreserved cells or tissues considering the viability and functionality is paramount to determine the quality and reliability of the cryopreservation protocol and solution. In the past, classical parameters, such as survival rate or motility, were the only quality measurements [22, 23]. With the evolution in technologies and developed assays, scientists can obtain more information surrounding the level of stress that heralds cellular death cascades and dynamic changes that impact cryopreserved cells' function and morphology.

Nowadays, there are a wide range of viability assays available; however, selecting the appropriate assay mainly depends on cell types to avoid inaccurate measurement. For instance, the measurement of LDH leakage in media can be used for membrane integrity assessment because of its reliability and easy performance. It is an applicable measurement in single cells as well as tissues and organs [24]. Conversely, using fluorescent probes for viability measurement is suitable for many cells excluding hepatocytes, because of their detoxification activity with respect to probes that influencing the accuracy of the measurement [25].

The emergence of developed technologies, such as genomics, transcriptomics, proteomics, and metabolomics (collectively termed OMICs), has provided a comprehensive profile of cryopreserved cells, including their stressed and compromised biological pathways, which may help designing protocols or solutions in order to modulate the damaged pathways. So far, the majority of OMICs applications in cryopreservation are limited to reproduction medicine and plants [26, 27], such as in human sperm characterization post-thaw [28]. The deep analysis OMICs stresses the importance of adopting such analytical approach in researches aiming at advancing cryopreservation and biobanking for better CTMPs outcome [25].

4. Cryoprotectant toxicity and detrimental effects

Introducing CPAs in high concentration (molars) is accompanied with non-specific adverse effects such as osmotic stress and cell dehydration [29] that also could induce the oxidative stress [30]. This can cause severe cell damage; for instance, increasing the concentration of DMSO, glycerol, and 1,2-propanediol is linked with the production of non-enzymatic formaldehyde [31], a cytotoxic compound that contributes to cell death [32]. The long exposure duration of cells to high concentration of CPA also harm cell development, as reported when exposing bovine blastocytes to a high concentration of ethylene glycol over 10 min [33]. Likewise, introducing a high concentration of propanediol to mouse zygotes was found to have a similar damaging effect on cell development to that observed in bovine blastocytes [34].

The high concentration of CPA accumulated intracellularly has a detrimental effect on cells. In cryopreserved human mesenchymal stem cells (hMSC), it has a significant effect on cellular viability, filamentous actin distribution, intracellular pH, and mitochondria aggregation [35]. It has also been found to cause abnormal spindles and morphology in human oocytes, which can potentially influence their viability post-cryopreservation [36]. Similarly, CPA causes a serious alteration in mammalian sperm viability, physiological properties, protein phosphorylation patterns [37], and can lethally damage enzymatic activity and DNA [38]. However, osmotic stress factors and associated cell shock cannot be decoupled since they interact with each other, though the resultant effects can be reversed or limited to a certain extent by minimizing exposure time, accelerating freezing and thawing speeds, and gradually diluting CPAs in cells [39], which can increase post-thaw cell viability. These types of reported damages are considered non-specific since it is not limited to specific CPA identity. However, the molecular interaction of CPAs is more closely linked to the permeable CPAs, as they are able to interact with the cell compartments and biomolecules [30].

CPA toxicity effect can be either reversible (e.g., osmotic shock and cellular shrinkage) [40, 41] or irreversible. Notably, cryopreservation protocols involving short exposure times to CPAs can reverse the induced damages. Nevertheless, irreversible damage is common in cells lacking self-renewal or repair mechanisms, such as RBCs [42] and embryonic stem cells [43, 44].

Oxidative stress occurs during cryopreservation, mainly when adding CPAs to cells [45]. The increased oxidative stress results in more ROS production [46], which leads to a disequilibrium between the generated ROS and the cellular antioxidant capacity within the redox pathway. A decrease in cellular-reduced glutathione (GSH) content was observed during the freezing step of sperm [47], indicating that oxidative damage occurs during the initial steps of cryopreservation. Consequently, increased ROS production results in lipid peroxidation [48], DNA instability [49], protein oxidation [50], overall dysfunctional cells, and low survival rates [47, 49]. Oxidative stress has been observed when applying glycerol [51], DMSO [52], and trehalose [50] to cells.

4.1 Other biochemical effects

Cells naturally have a dynamic and complex system involving active biomolecules that respond distinctly to all forms of environmental stressors, including CPA media and temperature alterations. The cells' response to stressors involves complex biomolecular events influencing their fate. Measuring the survival rate of thawed cells is a classical parameter that is not precise when determining the efficacy of cryopreservation. This is because during the recovery period, a decrease in cellular viability occurs in different cell types [53]. This is attributed to the activation of apoptosis machinery post-thaw [54]. Xu et al. [53] reported that exposing cells to DMSO and freezing conditions activate apoptosis through extrinsic and intrinsic pathways, including caspase-8, caspase-9, and p53. Some CPAs have different mechanisms, yet they lead to the same lethal results. Propylene glycol (ProH), for instance, reduced cell viability via increasing intracellular calcium to a cytotoxic level [55].

Furthermore, the cryopreservation affects cells' biomarkers [56]. It alters the proteome profile of cells, which in some cases can bring about changes in cellular metabolism, function, and structure [57]. In previous work, there is often no clear demarcation between the effect of CPAs and the cryopreservation protocol itself. However, the exact effect of CPAs can be investigated in an experiment if cell viability and functionality are analyzed before freezing.

5. Modulating CPA damages via additive agents

Considering the aforementioned limitations in cryomedia formula, many active studies investigated the efficacy of additive agents to improve the cryomedia and modulate the resultant damages in cryopreserved cells (**Table 1**). Additive agents have variable effects on different cells. This was evidently observed in number of cases such as; quercetin, glutathione, and ascorbic acid [58]. On other hand, some other demonstrated similar efficient antioxidant protection effect on several cells

Additive agents	Example	Concentration	Cell types	Molecular and biological effects
Antioxidants	Resveratrol [61] Salidroside [62]	15 μM 200 μM	Human sperms Red blood cells	Decrease DNA fragmentation through activating AMP-activated protein kinase (AMPK) Increase glutathione reductase (GR) activity and cells stability post thawing Reduce hemolysis, lactate dehydrogenase activity and protect protein and lipid from oxidation damage
Proteins	Type III anti-freezing proteins [63] Sericin [64]	0.8 mg/ml 5%	Human carcinoma cells Human sperm	Increase cells recovery post-thawing Increase cells motility Decreased DNA fragmentation
Enzymes	Catalase [65]	40 μl/ml	Mice spermatogonia stem cells	Reduce apoptosis and ROS production Increase viability
Vitamins	Vit E [66]	100–200 μmol	Human sperm	Increase motility
Anti-apoptotic drugs	Sphingosine-1- and Z-VAD-FMK [67]	10 μM	Ovarian sheep	Preserve primordial follicular density, with normal morphology and improved proliferation

Table 1.
Cryomedia additive agents and their effects on cryopreserved cells.

(e.g., curcumin) [58]. Notably, many protective factors share their antioxidant protection effects at different concentrations (e.g., hyaluronan and glutamine [59, 60]) that commonly include reducing oxidative stress on lipid and proteins and improve viability rate.

In our published studies, the discovery of the protection potent of salidroside and nigerose was exceptional on nucleated as well as anucleated hematopoietic cells [RBCs and human leukemia cells (HL-60)] in various cryomedia formulae and freezing modes. The efficacy of these compounds was evidenced at low concentrations (200–300 μM) of salidroside and nigerose, respectively. The effect of the additive compounds was determined by analyzing both the biomolecular and proteomic profiles of the survival cells [58]. First, we examined the effect of salidroside in standard cryo-solutions (glycerol and trehalose), which are commonly used for the RBCs biopreservation, using RBCs [62]. When comparing the survival cells rate, RBCs cryopreserved in solutions contained salidroside showed higher survival rate in compare to those cryopreserved in standard cryomedia alone. On biomolecular level, salidroside improved the intracellular activity of glutathione reductase (GR), the active enzyme in the redox pathway. In addition, it reduced the level of stress resultant from freeze-thaw process, as it was measured by intracellular lactate dehydrogenase (LDH) activity [68]. Moreover, it protected RBC proteins against oxidative damages [62]. Further investigation on human leukemia cells (HL-60) using salidroside in 2% DMSO and fetal bovine serum cryosolution demonstrated similar protection effects to what have been seen in RBCs [62, 68]. Additionally, it protected lipid against oxidative stress. In the same study, we used nigerose for comparison, which showed similar protection effect on the biomolecular profile of the cells.

On top of the biological profile of cryopreserved cells, proteomic analysis revealed the specific and unique modulation effect of additive agents on compromised biological pathways [68]. Each compound was observed to have a demonstrably unique effect on the proteome pattern of cryopreserved HL-60 cells. Nigerose was strongly engaged with cell maintenance, energetic, and metabolic pathways, whereas salidroside influenced proteins associated with DNA binding and nuclear activities. Both overlapped with regards to influencing proteins associated with redox pathways. Moreover, the damaging effects of classical cryomedia were modulated by the reformulated media comprising the novel protective agents. The protective mechanisms of the compounds on the proteomic level were strongly compatible with the biochemical analysis of the cells cryosurvival rate and their resistance to stressors [68]. This has shed the light over the potency of specific effectiveness of additive agents in the cryosolution and their specific applications for preserving different cells and tissues for pharmaceutical and clinical applications.

6. Conclusion

Understanding of the protective mechanisms of cryomedia ingredients along with identifying powerful protective compounds to enhance cryomedia performance is highly demanded. Due to the wide range of preserved cells and tissues, designing the appropriate cryosolution with suitable protocol is beneficial. In fact, these are particularly important for CTMP industries and end-users at clinics, such as those with cancer and diabetes or requiring blood transfusion, organ transplantation, and infertility treatments.

Author details

Noha A. Al-Otaibi
Life Science and Environment Research Institutes, King Abdulaziz City for Science and Technology, Riyadh, Saudi Arabia

*Address all correspondence to: naalotaibi@kacst.edu.sa

References

[1] Taylor MJ, Weegman BP, Baicu SC, Giwa SE. New approaches to cryopreservation of cells, tissues, and organs. Transfusion Medicine and Hemotherapy. 2019;**46**:197-215

[2] Wowk B. How cryoprotectants work. Cryonics. 2007;**3**:28

[3] Karow AM. Cryoprotectants—A new class of drugs. The Journal of Pharmacy and Pharmacology. 1969;**21**(4):209-223

[4] Hubálek Z. Protectants used in the cryopreservation of microorganisms. Cryobiology. 2003;**46**(3):205-229

[5] Lovelock JE. The mechanism of the protective action of glycerol against haemolysis by freezing and thawing. Biochimica et Biophysica Acta. 1953;**11**:28-36. Available at: http://www.sciencedirect.com/science/article/pii/0006300253900055

[6] Swain JE, Smith GD. Cryoprotectants. In: Chain R-C, Quinn P, editors. Fertility Cryopreservation. Cambridge: University of Cambridge Press; 2017. pp. 24-33

[7] Thanner M, Nagel E. A comprehensive assessment of ATMP. Difficulties and approaches. Bundesgesundheitsblatt - Gesundheitsforschung - Gesundheitsschutz. 2011;**54**(7):843-848. Available at: http://www.ncbi.nlm.nih.gov/pubmed/21698538

[8] Karlsson JOM, Younis AI, Chan AWS, Gould KG, Eroglu A. Permeability of the rhesus monkey oocyte membrane to water and common cryoprotectants. Molecular Reproduction and Development. 2009;**76**(4):321-333

[9] Seki A, Miyauchi S, Hayashi S, Kikukawa T, Kubo M, Demura M, et al. Heterologous expression of Pharaonis Halorhodopsin in Xenopus laevis oocytes and electrophysiological characterization of its light-driven Cl− pump activity. Biophysical Journal. 2007;**92**(7):2559-2569. Available at: http://linkinghub.elsevier.com/retrieve/pii/S0006349507710603

[10] Valdez DM, Miyamoto A, Hara T, Seki S, Kasai M, Edashige K. Water- and cryoprotectant-permeability of mature and immature oocytes in the medaka (Oryzias latipes). Cryobiology. 2005;**50**(1):93-102

[11] Pedro PB, Yokoyama E, Zhu SE, Yoshida N, Valdez DM, Tanaka M, et al. Permeability of mouse oocytes and embryos at various developmental stages to five cryoprotectants. The Journal of Reproduction and Development. 2005;**51**(2):235-246

[12] Mohanty JG, Nagababu E. Rifkind JM. Red blood cell oxidative stress impairs oxygen delivery and induces red blood cell aging. Frontiers in Physiology. 2014;**5**:84. Available at: http://www.pubmedcentral.nih.gov/articlerender.fcgi?artid=3937982&tool=pmcentrez&rendertype=abstract

[13] Crowe JH, Crowe LM, Wolkers WF, Oliver AE, Ma X, Auh J-H, et al. Stabilization of dry mammalian cells: Lessons from nature. Integrative and Comparative Biology. 2005;**45**(5):810-820. Available at: http://www.ncbi.nlm.nih.gov/pubmed/21676832

[14] Jain NK, Roy I. Effect of trehalose on protein structure. Protein Science. 2009;**18**:24-36

[15] Acker JP. The use of intracellular protectants in cell biopreservation. In: Baust JG, Baust JM, editors. Advances in Biopreservation. New york, Taylors and Francis; 2007. pp. 291-313

[16] Colaço C, Sen S, Thangavelu M, Pinder S, Roser B. Extraordinary stability of enzymes dried in trehalose:

Simplified molecular biology. Bio/ Technology. 1992;**10**(9):1007-1011

[17] Xie G, Timasheff SN. The Thermodynamic Mechanism of Protein Stabilization by Trehalose. Biophysical Chemistry. Elsevier; 1997. pp. 25-43

[18] Lynch AL, Chen R, NKH S. pH-responsive polymers for trehalose loading and desiccation protection of human red blood cells. Biomaterials. 2011;**32**(19):4443-4449. Available at: http://www.ncbi.nlm.nih.gov/pubmed/21421265

[19] Sampedro JG, Guerra G, Pardo J-P, Uribe S. Trehalose-mediated protection of the plasma membrane H+-ATPase from Kluyveromyces lactis during freeze-drying and rehydration. Cryobiology. 1998;**37**(2):131-138. Available at: http://www.sciencedirect.com/science/article/pii/S0011224098921095

[20] Radaelli MRM, Almodin CG, Minguetti-Câmara VC, Cerialli PMA, Nassif AE, Gonçalves AJ. A comparison between a new vitrification protocol and the slow freezing method in the cryopreservation of prepubertal testicular tissue. JBRA Assisted Reproduction. 2017;**21**(3):188-195

[21] Lee S, Ryu K-J, Kim B, Kang D, Kim YY, Kim T. Comparison between slow freezing and Vitrification for human ovarian tissue cryopreservation and xenotransplantation. International Journal of Molecular Sciences. 2019;**20**(13):3346

[22] Deller RC, Vatish M, Mitchell DA, Gibson MI. Glycerol-free cryopreservation of red blood cells enabled by ice-recrystallization-inhibiting polymers. ACS Biomaterials Science & Engineering. 2015;**1**(9):789-794. Available at: http://pubs.acs.org/doi/10.1021/acsbiomaterials.5b00162

[23] Clarke DM, Yadock DJ, Nicoud IB, Mathew AJ, Heimfeld S. Improved post-thaw recovery of peripheral blood stem/progenitor cells using a novel intracellular-like cryopreservation solution. Cytotherapy. 2009;**11**(4):472-479

[24] Drent M, NAM C, Henderson RF, EFM W, Van Dieijen-Visser M. Usefulness of lactate dehydrogenase and its isoenzymes as indicators of lung damage or inflammation. European Respiratory Journal. 1996;**9**:1736-1742

[25] Van BRG. Viability and functional assays used to assess preservation efficacy. In: Baust JG, Baust JM, editors. Advances in Biopreservation. Boca Raton: Taylor & Francis; 2007. pp. 123-141

[26] Egea R, Escrivá M, Puchalt N, Varghese A. OMICS: Current and future perspectives in reproductive medicine and technology. Journal of Human Reproductive Sciences. 2014;**7**(2):73. Available at: http://www.jhrsonline.org/text.asp?2014/7/2/73/138857

[27] Volk GM. Application of functional genomics and proteomics to plant cryopreservation. Current Genomics. 2010;**11**(1):24-29. Available at: http://www.pubmedcentral.nih.gov/articlerender.fcgi?artid=2851113&tool=pmcentrez&rendertype=abstract

[28] Wang S, Wang W, Xu Y, Tang M, Fang J, Sun H, et al. Proteomic characteristics of human sperm cryopreservation. Proteomics. 2014;**14**(2-3):298-310

[29] Arakawa T, Carpenter JF, Kita YA, Crowe JH. The basis for toxicity of certain cryoprotectants: A hypothesis. Cryobiology. 1990;**27**(4):401-415

[30] Elliott GD, Wang S, Fuller BJ. Cryoprotectants: A review of the actions and applications of cryoprotective solutes that modulate cell recovery from

ultra-low temperatures. Cryobiology. 2017;**76**:74-91

[31] Karran G, Legge M. Non-enzymatic formation of formaldehyde in mouse oocyte freezing mixtures. Human Reproduction. 1996;**11**(12):2681-2686. Available at: http://www.ncbi.nlm.nih.gov/pubmed/9021372

[32] Saito Y, Nishio K, Yoshida Y, Niki E. Cytotoxic effect of formaldehyde with free radicals via increment of cellular reactive oxygen species. Toxicology. 2005;**210**(2-3):235-245

[33] Sommerfeld V, Niemann H. Cryopreservation of bovine in vitro produced embryos using ethylene glycol in controlled freezing or vitrification. Cryobiology. 1999;**38**(2):95-105

[34] Damien M, Luciano AA, Peluso JJ. Propanediol alters intracellular pH and developmental potential of mouse zygotes independently of volume change. Human Reproduction. 1990;**5**(2):212-216

[35] Xu X, Liu Y, Cui Z, Wei Y, Zhang L. Effects of osmotic and cold shock on adherent human mesenchymal stem cells during cryopreservation. Journal of Biotechnology. 2012:224-231

[36] Mullen SF, Agca Y, Broermann DC, Jenkins CL, Johnson CA, Critser JK. The effect of osmotic stress on the metaphase II spindle of human oocytes, and the relevance to cryopreservation. Human Reproduction. 2004;**19**(5): 1148-1154

[37] Cole JA, Meyers SA. Osmotic stress stimulates phosphorylation and cellular expression of heat shock proteins in rhesus macaque sperm. Journal of Andrology. 2011;**32**(4):402-410

[38] Christoph K, Beck FX, Neuhofer W. Osmoadaptation of mammalian cells - an orchestrated network of protective genes. Current Genomics. 2007;**8**(4):209-218. Available at: http://www.ncbi.nlm.nih.gov/pubmed/18645598

[39] Meryman HT, Hornblower M. A method for freezing and washing red blood cells using a high glycerol concentration. Transfusion. 1972;**12**(3):145-156

[40] Ogura T, Shuba LM, McDonald TF. Action potentials, ionic currents and cell water in Guinea pig ventricular preparations exposed to dimethyl sulfoxide. Journal of Pharmacology and Experimental Therapeutics. 1995;**273**:1273-1286. Available at: http://www.ncbi.nlm.nih.gov/pubmed/7540688

[41] Pribor DB, Nara A. The effect of salt or various cryoprotective agents on frog sciatic nerves. Cryobiology. 1973;**10**(1):33-44

[42] Willekens FLA, Werre JM, Groenen-Döpp YAM, Roerdinkholder-Stoelwinder B, De Pauw B, Bosman GJCGM. Erythrocyte vesiculation: A self-protective mechanism? British Journal of Haematology. 2008;**141**(4):549-556

[43] Iwatani M, Ikegami K, Kremenska Y, Hattori N, Tanaka S, Yagi S, et al. Dimethyl Sulfoxide has an impact on epigenetic profile in mouse Embryoid body. Stem Cells. 2006;**24**(11):2549-2556. Available at: http://doi.wiley.com/10.1634/stemcells.2005-0427

[44] Tompkins JD, Hall C, Chen VC-Y, Li AX, Wu X, Hsu D, et al. Epigenetic stability, adaptability, and reversibility in human embryonic stem cells. Proceedings of the National Academy of Sciences of the United States of America. 2012;**109**(31):12544-12549. Available at: http://www.pnas.org/cgi/doi/10.1073/pnas.1209620109

[45] Yoon SJ, Rahman MS, Kwon WS, Park YJ, Pang MG. Addition of

cryoprotectant significantly alters the epididymal sperm proteome. PLoS One. 2016;**11**(3):1-15

[46] Chen G, Ren L, Zhang J, Reed BM, Zhang D, Shen X. Cryopreservation affects ROS-induced oxidative stress and antioxidant response in Arabidopsis seedlings. Cryobiology. 2015;**70**(1):38-47. Available at: http://linkinghub.elsevier.com/retrieve/pii/S0011224014007068

[47] Gadea J, Sellés E, Marco MA, Coy P, Matás C, Romar R, et al. Decrease in glutathione content in boar sperm after cryopreservation: Effect of the addition of reduced glutathione to the freezing and thawing extenders. Theriogenology. 2004;**62**(3-4):690-701

[48] Kadirvel G, Kumar S, Kumaresan A. Lipid peroxidation, mitochondrial membrane potential and DNA integrity of spermatozoa in relation to intracellular reactive oxygen species in liquid and frozen-thawed buffalo semen. Animal Reproduction Science. 2009;**114**(1-3):125-134

[49] Peris SI, Bilodeau JF, Dufour M, Bailey JL. Impact of cryopreservation and reactive oxygen species on DNA integrity, lipid peroxidation, and functional parameters in ram sperm. Molecular Reproduction and Development. 2007;**74**(7):878-892

[50] Kanias T, Acker JP. Trehalose loading into red blood cells is accompanied with hemoglobin oxidation and membrane lipid peroxidation. Cryobiology. 2015;**58**(2):232-239. Available at: http://www.ncbi.nlm.nih.gov/pubmed/19135990

[51] Baumber J, Ball BA, Linfor JJ, Meyers SA. Reactive oxygen species and cryopreservation promote DNA fragmentation in equine spermatozoa. Journal of Andrology. 2003;**24**(4):621-628

[52] Li P, Li Z-H, Dzyuba B, Hulak M, Rodina M, Linhart O. Evaluating the impacts of osmotic and oxidative stress on common carp (Cyprinus carpio, L.) sperm caused by cryopreservation Techniques1. Biology of Reproduction. 2010;**83**(5):852-858. Available at: https://academic.oup.com/biolreprod/article-lookup/doi/10.1095/biolreprod.110.085852

[53] Xu X, Cowley S, Flaim CJ, James W, Seymour L, Cui Z. The roles of apoptotic pathways in the low recovery rate after cryopreservation of dissociated human embryonic stem cells. Biotechnology Progress. 2010;**26**(3):827-837

[54] de Boer F, Dräger AM, Pinedo HM, Kessler FL, Monnee-van Muijen M, Weijers G, et al. Early apoptosis largely accounts for functional impairment of $CD34^+$ cells in frozen-thawed stem cell grafts. Journal of Hematotherapy & Stem Cell Research. 2002;**11**(6):951-963. Available at: http://online.liebertpub.com/doi/abs/10.1089/152581602321080619

[55] Takahashi T, Igarashi H, Doshida M, Takahashi K, Nakahara K, Tezuka N, et al. Lowering intracellular and extracellular calcium contents prevents cytotoxic effects of ethylene glycol-based Vitrification solution in unfertilized mouse oocytes. Molecular Reproduction and Development. 2004;**68**(2):250-258

[56] Kadić E, Moniz RJ, Huo Y, Chi A, Kariv I. Effect of cryopreservation on delineation of immune cell subpopulations in tumor specimens as determinated by multiparametric single cell mass cytometry analysis. BMC Immunology. 2017;**18**(1):6. https://doi.org/10.1186/S12865-017-0192-1

[57] Nynca J, Arnold GJ, Fröhlich T, Ciereszko A. Cryopreservation-induced alterations in protein composition of

rainbow trout semen. Proteomics. 2015;**15**(15):2643-2654

[58] Al-Otaibi NAS. Novel cryo-protective agents to improve the quality of cryopreserved mammalian cells [PhD Thesis]. University of Cambridge; 2018

[59] Ujihira M, Iwama A, Aoki M, Aoki K, Omaki S, Goto E, et al. Cryoprotective effect of low-molecular-weight hyaluronan on human dermal fibroblast monolayers. Cryo-Letters. 2010;**31**(2):101-111

[60] Takeo T, Sztein J, Nakagata N. The CARD method for mouse sperm cryopreservation and in vitro fertilization using frozen-thawed sperm. In: Methods in Molecular Biology. New York, NY: Humana Press; 2019. pp. 243-256

[61] Shabani Nashtaei M, Nekoonam S, Naji M, Bakhshalizadeh S, Amidi F. Cryoprotective effect of resveratrol on DNA damage and crucial human sperm messenger RNAs, possibly through 5′ AMP-activated protein kinase activation. Cell and Tissue Banking. 2018;**19**(1):87-95

[62] NAS A, NKH S, Rahmoune H. Salidroside as a novel protective agent to improve red blood cell cryopreservation. PLoS One. 2016;**11**(9):e0162748. Available at: http://www.pubmedcentral.nih.gov/articlerender.fcgi?artid=5025239&tool=pmcentrez&rendertype=abstract

[63] Tomás RMF, Bailey TL, Hasan M, Gibson MI. Extracellular antifreeze protein significantly enhances the cryopreservation of cell monolayers. Biomacromolecules; 14 Oct 2019;**20**(10):3864-3872

[64] Aghaz F, Khazaei M, Vaisi-Raygani A, Bakhtiyari M. Cryoprotective effect of sericin supplementation in freezing and thawing media on the outcome of cryopreservation in human sperm. The Aging Male. 2018. DOI: 10.1080/13685538.2018.1529156

[65] Aliakbari F, Sedighi Gilani MA, Yazdekhasti H, Koruji M, Asgari HR, Baazm M, et al. Effects of antioxidants, catalase and α-tocopherol on cell viability and oxidative stress variables in frozen-thawed mice spermatogonial stem cells. Artificial Cells, Nanomedicine, and Biotechnology. 2017;**45**(1):63-68

[66] Taylor K, Roberts P, Sanders K, Burton P. Effect of antioxidant supplementation of cryopreservation medium on post-thaw integrity of human spermatozoa. Reproductive Biomedicine Online. 2009;**18**(2):184-189

[67] Henry L, Fransolet M, Labied S, et al. Supplementation of transport and freezing media with anti-apoptotic drugs improves ovarian cortex survival. Journal of Ovarian Research. 2016;**9**:4. https://doi.org/10.1186/S13048-016-0216-0

[68] Al-Otaibi NAS, Cassoli JS, Martins-De-Souza D, Slater NKH, Rahmoune H. Human leukemia cells (HL-60) proteomic and biological signatures underpinning cryo-damage are differentially modulated by novel cryo-additives. GigaScience. 2018;**8**(3):1-13

Section 2

Procedures for Cryopreserving Gametes

Chapter 4

Cryopreservation of Human Spermatozoa: A New Frontier in Reproductive Medicine

Nabil Sayme

Abstract

Cryopreservation is a worldwide technique that makes it possible to preserve different living cells and tissues, including male and female gametes and embryos, in a structurally intact state using low temperature over time. Since the starting point of the cryopreservation era in 1776, until today, this was one of the most important steps in assisted reproductive techniques. Conventional slow freezing of spermatozoa is commonly used for cryopreservation of both ejaculated and surgically retrieved spermatozoa. The technique of the slow freezing is principally based on dehydration of cells which is performed through slow cooling combined with low concentrations of a cryoprotectant agent for achieving a balance. Besides of slow freezing, for more than a decade, many reports suggest the sperm vitrification technique as an alternative to slow freezing. Contrary to the slow freezing method, with vitrification, the effects of the cryoprotectants in spermatozoa are eliminated since this method is cryoprotectant-free. All of these interesting and promising protocols of vitrification, however, have not been implemented in the lab routine yet, and slow freezing remains the standard cryopreservation method in most laboratories worldwide.

Keywords: cryopreservation, vitrification, permeable and non-permeable cryoprotectants, human spermatozoa

1. Introduction

Cryopreservation, derived from the Greek word κρύος (krýos, "icy cold, chill, frost"), is a worldwide technique that makes it possible to preserve different living cells and tissues, including male and female gametes and embryos, in a structurally intact state using low temperature over time. It has become an indispensable part of most human-assisted reproductive technology (ART) programs around the world. Since the starting point of the cryopreservation era in 1776, Spallanzani [1] used snow to freeze spermatozoa and assess their motility after thawing. One century later, Mantegazza (reported by Curry [2]) observed that human sperm survived frozen at −17°C for more than 4 days. Since the late 1930–1940s [3, 4], scientists have effectively cryopreserved spermatozoa of several mammalian species, especially bovine and human. In 1947, glycerol was rediscovered as a cryoprotective agent allowing to freeze viable spermatozoa for longer periods. Bernstein and Petropavlovski [3] demonstrated the positive effect of 1 mol/l glycerol on spermatozoa frozen to −21°C in the rabbit, guinea pig, bull, ram, stallion, and boar. After that in the late 1940s, the results of experiments based on the use of glycerol by Polge et al. [4] in the United Kingdom (1949) were published. The first piglet born from frozen–thawed porcine sperm in 1957 showed that the fertilizing potential could also be preserved and no major harm to the genetic apparatus was done [5].

Some cryopreservation methods developed in the 1950s are in use till now. However, the real success of cryopreservation was achieved in the 1970s with the introduction of dry ice and liquid nitrogen vapor. Following the report published in 1964, for the first time, human spermatozoa were successfully cryopreserved [6]. Definitely, this was one of the most important steps in assisted reproductive techniques, since the preservation of male sample in some conditions is the only opportunity to preserve fertility for some couples. Besides that, sperm cryopreservation is also considered as a rescue option for cases prior to radiotherapy and/or chemotherapy [7] in cancer patients, prior to any other medical procedure that may potentially lead to testicular failure or ejaculatory dysfunction as vasectomy [8], as well as in cases of traveling husbands serving in the military or absent partners. Even in pediatric oncologic cases, in which testicular tissue can be preserved, the cryopreservation of immature testicular tissue for later fertilization purposes seems also advisable. Generally, sperm cryopreservation when correctly performed allows long-term storage and usage when needed, which is one of the most essential parts in assisted reproductive techniques.

Complete cryopreservation as a technique consists three steps: the first one is a collection of the sample, the second is the freezing procedure, and the third is the storage part in liquid nitrogen [9]. For the male gamete cryopreservation, collecting a sample can be divided into collecting ejaculated spermatozoa, collecting epididymal spermatozoa, and collecting testicular spermatozoa. The first one, collecting ejaculated spermatozoa, should be always the first option when the patient is capable to provide a sample with adequate sperm viability; since intracytoplasmic sperm injection (ICSI) allows that sperm is directly injected inside of the oocyte, minimum requirement for semen parameters do not exist. On the other hand, the complete absence of sperm in the ejaculate following 2–3 days of abstinence on at least two occasions is the standard used to confirm the diagnosis of azoospermia; in this case, epididymal and testicular collecting of spermatozoa are usual procedures [9, 10]. Percutaneous epididymal sperm aspiration (PESA) does not require a surgical incision. A small needle is going directly into the head of the epididymis through the scrotal skin, and fluid is aspirated and examined for the presence of motile sperm. On the other side, microsurgical epididymal sperm aspiration (MESA) involves dissection of the epididymis under the operating microscope and incision of a single tubule. Fluid spills from the epididymal tubule and then is aspirated. Epididymal spermatozoa are mature and progressively motile, and epididymal aspirates are much cleaner and devoid of the cellular debris that is seen in testicular sperm preparations [10]. For the testicular sperm extraction (TESE) procedures, different techniques have been developed and compared. The microsurgical TESE seems to have the highest sperm retrieval rate and may limit damage to the testicular tissue. Spermatozoa can be retrieved from tubules that are dilated, and this can be visualized with an operating microscope. This technique needs microsurgical skills and general anesthesia is usually required. An open biopsy also allows the excision of a larger tissue mass, allowing access to a greater number of sperm available for freezing [10, 11]. Besides these freezing steps, the thawing procedure is an equivalently important step. Spermatozoa must be allowed to retrieve its normal biological activities. Generally speaking, at the present time, all the cryopreservation protocols use a thawing temperature of 37°C for 10 or 15 min [12].

2. The conventional "slow freezing" method

Conventional slow freezing of spermatozoa is commonly used for cryopreservation of both ejaculated and surgically retrieved spermatozoa. The technique of slow freezing is principally based on dehydration of cells [13] which is performed through slow cooling combined with low concentrations of a cryoprotectant agent for achieving a balance. Slow freezing of spermatozoa is commonly performed by stepwise manual or continuous programmed freezing of vials or straws, containing a mixture of cryoprotectants and spermatozoa, to subzero temperatures [14]. The manual method is performed by simultaneously decreasing the temperature of the semen while adding a cryoprotectant in a stepwise manner and after immersing the samples into liquid nitrogen [15]. Mahadevan et al. [16] reported that the optimal initial cooling rate of the specimen from room temperature to 5°C is 0.5–1°C/min. The sample is then frozen from 5°C to −80°C at a rate of 1–10°C/min and then submerged into liquid nitrogen at −196°C. Despite the fact many research presented successful sperm freezing with manual techniques, the reproducibility of this procedure pointed out some problems. For this reason, programmable freezers have been investigated [17]. These devices generally use a plate to hold the straws; these are cooled by liquid nitrogen held in a storage tank under the plate. Liquid nitrogen is poured into the tank, and the machine, once programmed, uses the software data logging to obtain cooling from 20°C to −80°C at a rate of 1.5°C/min and then at 6°C/min; at completion of the freezing, the straws are removed and stored in liquid nitrogen at −196°C. This takes about 40 min [17]. Software is simple to use and do not require continuous operator intervention and, besides that, have been used to increase the reproducibility of the freezing operations.

However, many research confirmed that the cryopreserved/thawed spermatozoa lose about 50–40% of their pre-freezing motility value, with considerable fluctuation among samples [18]. Although, after freezing, the regression of the motility is one of the first affected parameters, the mechanism of sperm impairment and its mechanical, physical, and/or chemical etiology is still the point of discussion. The ice equilibrium in conventional slow freezing is one of the main causes of mechanical cell injury, which result in the formation of intra- or extracellular ice crystal, along with the osmotic damage [19].

The cell further downgrades as well through consequent thawing and rewarming that affect their viability by possible excessive osmotic swelling. Chemical and physical damage are also caused affecting the sperm cell membranes as a result of expanded lipid peroxidation, due to the production of reactive oxygen species [20] which also lead to loss of sperm motility [21]. In order to prevent all the previously reported damages, during the time, different cryoprotectants were developed. Cryoprotectants are low-molecular-weight and highly permeable chemicals used to protect spermatozoa from freeze damage by ice crystallization. There are four main well-known cryoprotectants: glycerol, ethylene glycol, dimethyl sulfoxide, and 1,2-propanediol. Cryoprotectants decrease the freezing point of a sample, reduce the amount of salts and solutes present in the liquid phase of the sample, and decrease ice formation within the spermatozoa [22]. Usually, the cryoprotectants are added in an equal volume of semen in a dropwise manner, gently mixed at room temperature, and then placed at 37°C for 10–15 min to allow for proper equilibration between the cells and the medium [23]. Besides the permeable cryoprotectants, there are non-permeable agents, such as raffinose, sucrose, egg yolk citrate, albumin, polyethylene glycol, and polyvinylpyrrolidone, which are common additives that cannot pass through the plasma membrane but have antioxidant effects and are used to improve post-thaw sperm functional parameters such as post-thaw motility, viability, and reducing DNA damage [12]. Presently, slow freezing techniques have been widely used in sperm cryopreservation with acceptable results for sperm vitality and motility after thawing [24]. However, for more than a decade, many reports suggest sperm vitrification technique as an alternative to slow freezing [25].

3. Vitrification as an alternative: new technique

In the past, the first successful vitrification of frog spermatozoa was done by Luyet and Hodapp [26] and fowl spermatozoa 4 years later by Shaffner [27]. In the early 1980s, Rall and Fahy [28] managed to successfully vitrify embryos using high concentrations of permeable cryoprotective agents and a relatively low speed of cooling and warming, and since then, the main approach to vitrifying spermatozoa has been considered the same as the methods used for other types of mammalian cells [29]. Vitrification as a method is based on the ultrarapid increase and decrease of temperatures with or without the use of non-permeable cryoprotectants. During the procedure, water is cooled to a glassy state through extreme increasing of viscosity without intracellular ice crystallization making this procedure less labor-intensive, faster, and presumably safer than traditional slow freezing protocols [30].

Contrary to the slow freezing method, the effects of the cryoprotectants in spermatozoa are eliminated since this method is cryoprotectant-free. Using this method, the sperm suspension is plunged directly into liquid nitrogen, and the sperm cells are cooled in an ultrarapid manner, known as kinetic vitrification [25]. What is more, cryoprotectant-free technique avoids the use of the classic toxic cryoprotective agents (CPA) that may have lethal effects of osmotic shock and prevent lethal intracellular ice formation and the harmful effects of high salt concentrations during freezing and thawing [25]. The first described cryoprotectant-free vitrification by Nawroth et al. [31] suggested that spermatozoa were located onto copper loops or into standard 0.25 ml insemination straws and plunged directly into liquid nitrogen. During the years of improving technique, different devices have been tested [30, 32, 33]. Isachenko et al. [34] compared the vapor phase and liquid phase for sperm cryopreservation, using the cryoloop, droplet, and open straw methods. Cryoloops with a film of spermatozoa suspension were cooled for 3 min in liquid nitrogen vapor at −160°C and then placed into pre-cooled cryovials (CryoTubesTM, 4.5 ml volume, 92 mm length; Nunc GmbH & Co. KG, Wiesbaden, Germany) and were stored in liquid nitrogen until the time of warming. The second developed method at that time was vitrification in liquid nitrogen vapor using droplets where sperm suspension was located onto aluminum foil previously cooled in liquid nitrogen vapor to −160°C. During cooling, the droplet of sperm suspension adopted a spherical form, which was placed into pre-cooled cryovials and stored in liquid nitrogen. With open straw method, sperm suspension was drawn inside the end of open-pulled straws (0.25 ml) (Medical Technology GmbH, Altdorf, Germany) by capillary action [32]. Straws were placed inside other sterile 90 mm straws which were prepared from the standard 0.5 ml insemination straws (Medical Technology GmbH, Altdorf, Germany) and then hermetically closed using a handheld sealer and then plunged into liquid nitrogen. During this procedure, it was described that there was no contact between the wall of the 90 mm straw and the suspension of spermatozoa inside the open-pulled straws, due to the presence of a meniscus in the suspension [34]. At the end, the results report that CPA-free cryopreservation of spermatozoa could occur in a wide range of cooling rates, but the major disadvantage of cryoprotectant-free vitrification was that only small volumes of spermatozoa (≤40 μl) could be vitrified in these systems [32].

What is more, in the aim to improve the cryoprotectant-free vitrification, Schulz et al. [35] added some carbohydrate supplements (glucose, sucrose, and trehalose) to the sperm suspension and obtained that sucrose has increased spermatozoa motility and viability after thawing. Later on, Isachenko et al. [36] investigates the ability of sucrose to protect sperm motility, viability, mitochondrial membrane potential integrity, spontaneous capacitation, and acrosome reaction. Spermatozoa were cryoprotectant-free vitrified using three different media: human tubal fluid medium (control), human tubal fluid medium with 1% human serum albumin, and fluid medium with 1% human serum albumin and 0.25 M sucrose [36].

Obtained results in this research [36] indicate that the number of progressively motile spermatozoa was significantly higher in the sucrose-supplemented medium group (57%) than with controls (19%). The combination of fluid medium with 1% human serum albumin and sucrose (65%) has a stronger cryoprotective effect on the integrity of mitochondrial membrane potential than with human tubal fluid medium with 1% human serum albumin alone (33%). It was concluded that cryoprotectant-free vitrification of human spermatozoa with non-permeable cryoprotectants such as human serum albumin and sucrose can effectively cryopreserve the cells without significant loss of important physiological parameters. On the other hand, Chen et al. [37] reported that normozoospermic patient samples vitrified by the cryoprotectant-free method with or without the addition of sucrose did not show a significant difference in the sperm recovery rate and motility rate.

During the years, many researchers got their attention into the vitrification of oocytes, embryos, larger cells, and even tissues [38–40]. It was established, at that time, that vitrification for oocyte and embryo could not be achieved without combinations of high concentrations of both permeable and non-permeable cryoprotectants in order to reach stable vitrification and allow using a relatively low rate of cooling and warming. Mainly dimethyl sulfoxide, propanediol, or ethylene glycol were used for ooctyte and embryo vitrification [41–43]. However, these methods also have their limitations, and especially human spermatozoa are intolerant to the high concentrations of cryoprotectants conventionally used in vitrification [44, 45]. Suggested reasons for successful vitrification in the absence of cryoprotectants for spermatozoa are the size of the cells and their relative concentration of soluble macromolecules. The shape and size of the sperm head could define the cryo-sensitivity of the cell. Comparative studies [46] have shown a negative correlation between the size of the sperm head and cryo-stability. Oocyte and spermatozoa naturally contain high concentrations of proteins, which help in vitrification. Thus, a higher cryoprotectant concentration is needed for extracellular vitrification than for intracellular vitrification [25]. It can be assumed that the amount of osmotically inactive water is also higher in spermatozoa and is combined with several macromolecular structures such as DNA, histones, etc. [47]. Extensive classification of intracellular compounds may also contribute to the successful survival of spermatozoa [48].

In spite of this, vitrification of spermatozoa is still a rather unexplored methodology, with limited studies showing its efficacy in male gametes [24, 25, 30–32]. In the last two decades, different sperm vitrification protocols have been published, most of them developed by Isachenko [30–32, 34, 36] where not only different combinations of cryoprotection but also different devices have been tested as well. These interesting and promising protocols, however, have not been implemented in the IVF lab routine yet, and slow freezing remains the standard cryopreservation method in most laboratories worldwide [49].

4. Our experience

Since there is no optimal accepted algorithm for the vitrification procedure and no evidence has been established regarding the possibility of achieving successful human spermatozoa vitrification without cryoprotectant, the differences between vitrification of human sperm with and without two non-permeable cryoprotectants (NPC), sucrose (SUC), and trehalose (TRE) were investigated. For that, five human semen samples diagnosed as normal (*normozoospermia)* were obtained and analyzed. All samples were prepared through 80% one-step density gradient and then were diluted to a defined sperm concentration of 40×10^6 sperms per ml. following the guidelines of the World Health Organization criteria [13]. Motility and vitality of the sperm were recorded in each sample. Each sample was then divided into three parts. Each part of the samples underwent vitrification using three different methods: (i) 150 μl cryostraw filled with the sample was directly plunged into liquid nitrogen, (ii) small drops of the samples were dripped into liquid nitrogen, and (iii) microcapillaries were plunged directly into liquid nitrogen. Each part was vitrified differently using sucrose and trehalose or without a cryoprotectant. Sucrose and trehalose were diluted in media, 500 mM, after merging it with the sperm solution concentrations till they have reached 250 mM of concentration with 20×10^6 sperms per ml. After a minimum of 24 hours of cryopreservation, samples were thawed in warm water at 37°C for 20 seconds. Then, the samples were washed in 1 ml media (SpermActive®; Gynemed) and incubated at 37°C. Motility and vitality were directly assessed, 30 and 120 min after thawing. It was found that sperms' overall survival rates ranged from 1.2 to 37.5%. Mean survival rates for each method were as follows: method (i), 15.5 ± 5.3% for sucrose, 16.6 ± 5.2% for trehalose, and 19.8 ± 7.8% without cryoprotectant; method (ii), 47.8 ± 16.6% for sucrose, 53.8 ± 13.4% for trehalose, and 40.4 ± 7.5 without cryoprotectant; and method (iii), 10.3 ± 2.9% for sucrose, 8.2 ± 3.9% for trehalose, and 8.2 ± 3.3% without cryoprotectant (**Table 1**). Statistical analyses revealed significant differences only between method (ii) and method (iii) (**Table 2**), with a tendency for better survival rates in method (ii) under all three conditions, especially with trehalose (mean survival rate of 53.8 ± 13.4%; **Table 1**). However, these differences are not representative because of the low survival and motility rate after thawing for method (iii), which might be affected by the heat sealing process (sealing small volume of the sample). No significant differences were observed regarding the addition of sucrose and trehalose or vitrification without a cryoprotectant in terms of overall survival rates (suc. vs. no cryoprotectant $n = 5$, $p > 0.4$; tre. vs. no cryoprotectant $n = 5$, $p > 0.3$; **Table 2**). Similarly, no significant differences were assessed between the use of sucrose and trehalose ($n = 5$, $p > 0.7$). Mean motility recovery, measured after 120 min of thawing, was 13.7 ± 5.9% for sucrose, 12.8 ± 4.2% for trehalose, and 10.1 ± 4.3% without cryoprotectant, and

	N	Mean Survival
Method (i) + sucrose	5	15.6 ± 11.7%
Method (i) + trehalose	5	16.6 ± 11.6%
Method (i) without NPC	5	19.8 ± 17.5%
Method (ii) + sucrose	5	47.8 ± 16.6%
Method (ii) + trehalose	5	53.8 ± 13.4%
Method (ii) without NPC	5	40.4 ± 7.5%
Method (iii) + sucrose	5	10.3 ± 2.9%
Method (iii) + trehalose	5	8.2 ± 3.9%
Method (iii) without NPC	5	8.2 ± 3.3%

Table 1.
Mean survival rates for all three methods and conditions.

	Sucrose	Trehalose	without NPC
(i) vs. (ii)	P=0.207	P=0.217	P=0.131
(i) vs. (iii)	P=0.38	P=0.171	P=0.194
(ii) vs. (iii)	P=0.007	P=0.002	P=0.002

Table 2.
Comparison of the different methods.

between these three groups, there was not a significant difference. However, interestingly, a tendency of improved motility was observed during post-thawing incubation specifically in addition of cryoprotectants (**Figure 1**). Therefore, the choice of the vitrification method and conditions seems to influence survival rates, motility, and vitality of the sperm, but the significant difference in sperm recovery after vitrification with non-permeable cryoprotectants was not found; the reason

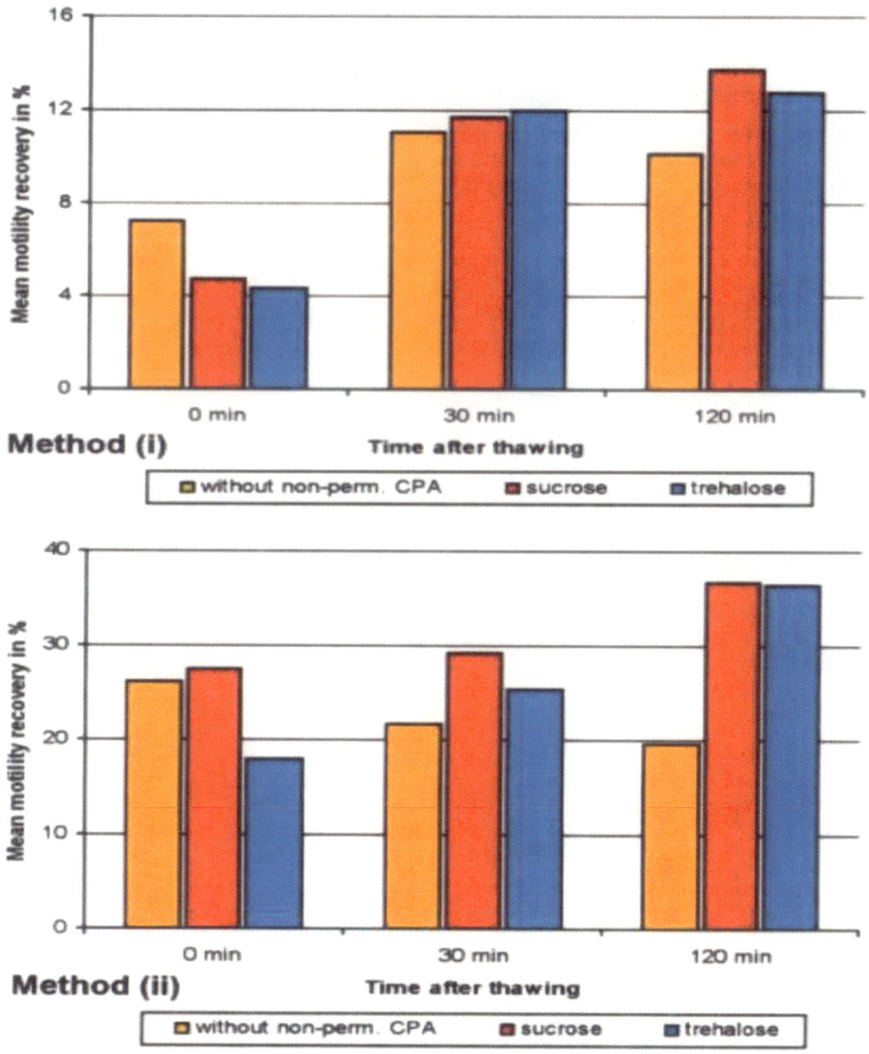

Figure 1.
Recovery of motility by post-thawing incubation.

that influenced the obtained result might be the small sample size. On the other hand, the improvement of the methodology of sperm vitrification could yield in positive additional effects of non-permeable cryoprotectants. Under these conditions, obtained data might be encouraging for further studies, to extend on a greater number of normal sperm samples as well as to those patients with reduced semen quality and fertility problems. Moreover, the increase in post-thaw sperm motility is an important aspect in the use of all assisted reproduction techniques and should be also confirmed by further studies.

5. Conclusion

Spermatozoa cryopreservation for males is the standard fertility preservation care in patients undertaking gonadotoxic treatments, such as chemotherapy/radiotherapy. The conventional cryopreservation, the slow freezing technique, is standardized and commonly used. However, functional sperm parameters including motility after the thawing are still challenging. New methods that preserve spermatozoa are promising, even though they still need validation before being routinely used in an assisted reproduction program, including the essential use of new cryoprotectants and new antioxidants to improve sperm quality after thawing. Moreover, spermatozoa lyophilization is another method that is still under investigation. However, as the spermatozoa are immotile, lyophilized sperm can only be used in ICSI. Hence, future research needs not only to investigate the optimization and safety of methods but also for the health of the offspring.

Author details

Nabil Sayme
Team Kinderwunsch Hannover, Hannover, Germany

*Address all correspondence to: nsayme5@aol.com

References

[1] Spallanzani L. Osservazioni e spezienze interno ai vermicellispermatici dell' uomo e degli animali. Modena: Opusculi di Fisica Animalee Vegetabile; 1776

[2] Curry MR. Cryopreservation of semen from domestic livestock. Day JG, McLellan MR, editors. Methods in Molecular Biology. Cryopreservation and Freeze-Drying Protocols. Vol. 38. New York: Humana Press Inc; 1995. pp. 189-197

[3] Bernstein AD, Petropavlovski VV. Influence of non-electrolytes on viability of spermatozoa. Buleten Experimentalnoi Biologii i Medicini. 1937;**III**:17-28

[4] Polge C, Smith AU, Parkes AS. Revival of spermatozoa after Vitrification and dehydration at low temperatures. Nature. 1949;**164**(4172):666. Available from: https://doi.org/10.1038/164666a0

[5] Hess EA, Teague HS, Ludwick TM, Martig RC. Swine can be bred with frozen semen. Ohio Fm Home Res. 1957;**42**:100. Available from: https://eurekamag.com/research/014/216/014216902.php

[6] Sherman JK. Synopsis of the use of frozen human semen since 1964: State of the art of human semen banking. Fertility and Sterility. 1973;**24**:397-416

[7] Sanger WG, Olson JH, Sherman JK. Semen cryobanking for men with cancer-criteria change. Fertility and Sterility. 1992;**58**:1024-1027

[8] Curry RM. Cryopreservation of semen from domestic livestock. Reviews of Reproduction. 2000;**1**(5):46-52

[9] Martins AD, Agarwal A, Henkel R. Sperm cryopreservation. In: Zsolt N, Varghese P, Alex C, Ashok A, editors. Vitro Fertilization. Switzerland: Springer International Publishing; 2019. pp. 625-639. Available from: https://doi.org/10.1007/978-3-319-43011-9_51

[10] Gangrade BK. Cryopreservation of testicular and epididymal sperm: Techniques and clinical outcomes in assisted conception. Clinics. 2013;**68**(S1):131-140

[11] Dohle GR, Elzanaty S, van Casteren NJ. Testicular biopsy: Clinical practice and interpretation. Asian Journal of Andrology. 2012;**14**:88-93

[12] Di Santo M, Tarozzi N, Nadalini wM, Borini A. Human sperm cryopreservation: Update on techniques, effect on DNA integrity and implications for art. Advances in Urology. 2012;**2012**:837-854

[13] Thachil JV, Jewett MAS. Preservation techniques for human semen. Fertility and Sterility. 1921;**35**(5):546-548

[14] World Health Organization. WHO Laboratory Manual for the Examination and Processing of Human Semen. 5th ed. Geneva: World Health Organization Press; 2010

[15] Said TM, Gaglani A, Agarwal A. Implication of apoptosis in sperm cryoinjury. Reproductive Biomedicine Online. 2010;**21**(4):456-462

[16] Mahadevan M, Trounson AO. Effect of cooling, freezing and thawing rates and storage conditions on preservation of human spermatozoa. Andrologia. 1984;**16**(1):52-60

[17] Holt WV. Basic aspects of frozen storage of semen. Animal Reproduction Science. 2000;**62**(1-3):3-22

[18] Keel BA, Webster BW, Roberts DK. Semen cryopreservation methodology

and results. In: Barratt CLR, Cooke ID, editors. Donor Insemination. Cambridge, UK: Cambridge University Press; 1993. pp. 71-96

[19] Maettner R, Sterzik K, Isachenko V, Strehler E, Rahimi G, Alabart JL, et al. Quality of human spermatozoa: Relationship between high-magnification sperm morphology and DNA integrity. Andrologia. 2013;**46**:547-555

[20] Hammerstedt RH, Graham JK, Nolan JP. Cryopreservation of mammalian sperm: What we ask them to survive. Journal of Andrology. 1990;**11**:73-88

[21] Baust JG, Gao D, Baust JM. Cryopreservation: An emerging paradigm change. Organogenesis. 2009;**5**:90-96

[22] Royere D, Barthelemy C, Hamamah S, Lansac J. Cryopreservation of spermatozoa: A 1996 review. Human Reproduction Update. 1996;**2**(6):553-559

[23] Fabbri R, Ciotti P, Di Tommaso B. Tecniche di crioconservazione riproduttiva. Giornale Italiano di Ostetricia e Ginecologia. 2004;**3**:33-41

[24] Riva NS, Ruhlmann C, Iaizzo RS, López CAM, Martínez AG. Comparative analysis between slow freezing and ultra-rapid freezing for human sperm cryopreservation. JBRA Assisted Reproduction. 2018;**22**(4):331

[25] Isachenko E, Isachenko V, Katkov II, Dessole S, Nawroth F. Vitrification of mammalian spermatozoa in the absence of cryoprotectants: From past practical difficulties to present success. Reproductive Biomedicine Online. 2003;**6**:191-200

[26] Luyet BJ, Hoddap A. Revival of frog's spermatozoa vitrified in liquid air. Proceedings of the Meeting of Society for Experimental Biology. 1938;**39**:433-434

[27] Schaffner CS. Longevity of fowl spermatozoa in frozen condition. Science. 1942;**96**:337

[28] Rall WF, Fahy GM. Ice-free cryopreservation of mouse embryos at −196°C by vitrification. Nature. 1985;**313**:573-575

[29] Rall WF. Prospects of the cryopreservation of mammalian spermatozoa by vitrification. In: Johnson LA, Katz D, editors. Boar Semen Preservation. Berlin-Hamburg: Paurey Scientific Publishers; 1991. pp. 125-156

[30] Katkov II, Isachenko V, Isachenko E, Kim MS, Lulat AG, Mackay AM, et al. Low- and high-temperature vitrification as a new approach to biostabilization of reproductive and progenitor cells. International Journal of Refrigeration. 2006;**29**:346-357

[31] Nawroth F, Isachenko V, Dessole S, Rahimi G, Farina M, Vargiu N, et al. Vitrification of human spermatozoa without cryoprotectants. Cryo Letters. 2002;**23**:93-102

[32] Isachenko V, Isachenko E, Katkov II, Montag M, Dessole S, Nawroth F, et al. Cryoprotectant free cryopreservation of human spermatozoa by vitrification and freezing in vapor: Effect on motility, DNA integrity, and fertilization ability. Biology of Reproduction. 2004b;**71**:1167-1173

[33] Vajta G, Booth PJ, Holm PCH. Successful vitrification of early stage bovine in vitro produced embryos with the open pulled straw (OPS) method. Cryo Letters. 1997;**18**:191-195

[34] Isachenko V, Isachenko E, Montag M, Zaeva V, Krivokharchenko A, Nawroth F, et al. Clean technique for cryoprotectant free vitrification of human spermatozoa. Reproductive Biomedicine Online. 2005;**10**:350-354

[35] Schulz M, Muñoz M, Risopatrón J, Sánchez R. Cryopreservation of human spermatozoa by vitrification. International Journal of Morphology. 2006;**24**:31

[36] Isachenko E, Isachenko V, Weiss JM, Kreienberg R, Katkov II, Schulz M. Acrosomal status and mitochondrial activity of human spermatozoa vitrified with sucrose. Reproduction. 2008;**136**:167-173

[37] Chen Y, Li L, Qian Y, Xu C, Zhu Y, Huang H, et al. Small-volume vitrification for human spermatozoa in the absence of cryoprotectants by using Cryotop. Andrologia. 2015;**47**:694-699

[38] Tucker M, Wright G, Morton P, Shanguo L, Massey J, Kort H. Preliminary experience with human oocyte cryopreservation using 1,2-propanediol and sucrose. Human Reproduction. 1996;**11**:1513-1515

[39] Hong SW, Chung HM, Lim JM, Ko JJ, Yoon TK, Yee B, et al. Improved human oocyte development after vitrification: A comparison of thawing methods. Fertility and Sterility. 1999;**72**:142-146

[40] Katayama KP, Stehlik J, Kuwayama M. High survival rate of vitrified human oocytes results in clinical pregnancy. Fertility and Sterility. 2003;**80**:223-224

[41] Ali J, Shelton JN. Design of vitrification solutions for the cryopreservation of embryos. Journal of Reproduction and Fertility. 1993;**99**:471-477

[42] Mukaida T, Wada M, Takahashi K. Vitrification of human embryos based on the assessment of suitable conditions for 8-cell mouse embryos. Human Reproduction. 1998;**13**:2874-2879

[43] Vajta G, Nagy ZP. Are programmable freezers still needed in the embryo laboratory? Review on vitrification. Reproductive Biomedicine Online. 2006;**12**:779-796

[44] Katkov II. The point of maximum cell water volume excursion in case of presence of an impermeable solute. Cryobiology. 2002a;**44**:193-203

[45] Katkov II. Symmetrical relativistic permeability approach simplifies optimization of cryobiological protocols. In: Proceedings of the International Institute of Refrigeration (IIR) International Conference, Joined Meeting of the IIR with the Society for low Temperature Biology: Cryopreservation and Safe Keeping of Cells and Tissues. Hradec Kralove, Czech Republic, May 13-1. 2002b

[46] Nauk VA. Structure and function of spermatozoa from farm animal by cryopreservation. Shtiinca Kishineu;1991 p. 198

[47] Isachenko E, Isachenko V, Katkov II, Rahimi G, Schöndorf T, Mallmann P, et al. DNA integrity and motility of human spermatozoa after standard slow freezing versus cryoprotectant-free vitrification. Human Reproduction. 2004;**19**:932-939

[48] Gao D, Mazur P, Critser J. Fundamental cryobiology of mammalian spermatozoa. In: Karow AM, Critser JK, editors. Reproductive Tissue Banking. London: Academic Press; 1997. pp. 263-328

[49] Mocé E, Fajardo AJ, Graham JK. Human sperm cryopreservation. EMJ. 2016;**1**:86-91

Chapter 5

Cryobiology and Cryopreservation of Sperm

Ali Erdem Öztürk, Mustafa Numan Bucak, Mustafa Bodu, Nuri Başpınar, İlhami Çelik, Zhiquan Shu, Nazan Keskin and Dayong Gao

Abstract

Low temperature has been utilized to keep living cells and tissues dormant but potentially alive for cryopreservation and biobanking with great impacts on scientific and biomedical applications. However, there is a critical contradiction between the purpose of the cryopreservation and experimental findings: the cryopreserved cells and tissues can be fatally damaged by the cryopreservation process itself. Contrary to popular belief, the challenge to the life of living cells and tissues during the cryopreservation is not their ability to endure storage at cryogenic temperatures (below −190°C); rather it is the lethality associated with mass and energy transport within an intermediate zone of low temperature (−15 to −130°C) that a cell must traverse twice, once during cooling and once during warming. This chapter will focus on (1) the mechanisms of cryoinjury and cryopretection of human sperm in cryopreservation, and (2) cryopreservation techniques and methods developed based on the understanding of the above mechanisms.

Keywords: cryopreservation, cryobiological characteristics, sperm

1. Introduction

The pioneering studies in cryobiological science starting from the middle of the last century is based on the idea that whether life span and longevity could be extended by storing various frozen tissues and organs or even human body for many years. Possible cellular and tissue destruction during the freeze-thaw processes became the subject of curiosity.

The male reproductive cell, spermatozoon, was first discovered in 1677 by Van Leeuwenhoek and was called "Animalacula." The early studies to obtain pregnancy by artificial insemination were done by Lazzaro Spallanzani in 1776 [1, 2]. The studies with scientific basis and modern vital cell freezing studies began in 1949 after the discovery of glycerol as a cryoprotectant by Polge et al. [3] and spermatozoon became the first mammalian cell to be successfully frozen [4]. Following the success in freezing spermatozoa, Whittingham et al. [5] successfully froze the mouse embryo by using dimethylsulfoxide (DMSO) as a cryoprotectant, in 1972. Nowadays, millions of children have been born from cryopreserved sperms or embroys.

Sperm cryopreservation has been successfully applied in various fields to benefit the human kind and animals, including assisted reproduction, rescuing the endangered species, and saving the ecosystem. However, we also know that damage can be caused to the cells during the cryopreservation process. In this review, the mechanisms of damage to the sperm cells during the process of cryopreservation will be taken under spotlight and we will try to elucidate them in a cause-effect manner.

2. What is cryobiology?

Cryobiology is a multidisciplinary science, studying the physical and biological behaviors of living materials (e.g., cells and tissues) at low temperatures. Cryobiology contains many disciplines such as, cellular biology, theriogenology and molecular biology, engineering and mathematics, veterinary and human medicine, intensive and extensive farming on land and in watery environments [6]. Optimization of the cryopreservation procedure of spermatozoa needs all the above-mentioned disciplines because of the complex cellular structure, activation and capacitation mechanisms of spermatozoon [7].

3. Dormant or dead? two aspects of freezing of cells

Freezing biological time occurs when the cells are cooled down in controlled manner under temperatures lower than necessary for continuing their normal physiological activities. Damaged or dying cells exhibit characteristic changes that cause structural differentiation under destruction process and are dragged into a possible death. In the present case, there are two types of scenarios, where cells may be damaged or killed: necrosis and apoptosis.

3.1 Necrosis

The cell death takes place through one of the pathways such as necrosis, necrobiosis, or apoptosis. In necrosis, a pathological cell death, the cell has been severely damaged as a result of sudden and extreme trauma, and soon death occurs. Because the cell loses control of permeability of plasmalemma, the cytosol rapidly swells by excessive inflow of water and ions, and organelle becomes excessively swollen. As a result, the cell lyses without spending energy. The cell contents and endotoxins leak out of the collapsed or inflammatory cells into the area where inflammatory changes begin. In this case, neighboring cells and tissue elements may also be exposed to damage as well [8]. Factors causing damage by disrupting the local homeostatic balance of the cells and tissues such as hypoxia, viral, and bacterial infections, toxins, radiation and changes in temperature are the main causes of necrosis. Necrosis, as an irreversible process, is also defined as a transition of cell into a definite death and loss of all of its physiological functions.

3.2 Apoptosis

Apoptosis, which is a programmed cell death differs from necrosis with a variety of morphological, biochemical, and physical changes.

Essentially, apoptosis is a physiological cell death, a natural process that occurs during embryonic development and periodically in organs such as mammary gland and uterus, which undergo cyclic changes. In this case, the cell systematically ceases all of its functions and breaks down by incorporating its structural components into vesicles [9]. The main histological differences between apoptosis and necrosis are summarized in **Table** 1.

Cellular results	Necrosis	Apoptosis
General cellular changes	The cell is swollen, massive cell death occurs in a very large area.	Cells die in groups, apoptotic bodies are formed.
Organelles	Damaged	Not damaged
Mitochondria	Mitochondria, due to lack of ATP swell and break down	Mitochondria are swollen, cytochrome-C is released.
Cell membrane	There is structural deterioration, selective permeability control is lost.	It is intact, the surface takes a crater appearance, vesicles are formed.
Nucleus	Chromatin loses its normal organization and is in the form of thick chromatin yarns. Pyknosis, karyorrhexis, or karyolysis develops.	Chromatin is fragmented, it is concentrated in the form of a hat, nucleolus is dispersed
Causal factors	Anoxia, starving, physical and chemical traumas that lead to ATP deficiency.	They are physiological and pathological conditions that do not lead to ATP deficiency.
Effects on the tissue	The inflammatory events develop, degenerated cell debris is found in the environment and these are engulfed by phagocytes. There is a common tissue destruction.	There is no inflammation, the resulting apoptotic vesicles are phagocytosed by neighboring cells and macrophages, a rapid involution occurs without collapse in the tissue.

Table 1.
Comparison of the effects of necrosis and apoptosis on cells, organelles, and tissues [10, 11].

Apoptosis consists of three stages as, final decision, execution, and cleaning phase. The cell receives a nonreversible lethal apoptotic stimulation beyond the time of decision. In the execution phase, condensation of chromatin in the cell, shrinking of the cytoplasm, formation of buds in the cell membrane, fragmentation of DNA, and formation of apoptotic bodies occur.

These changes are affected by different enzymes such as proteases, lipases, and nucleases. In the final stage of the cleaning process, apoptotic bodies are phagocytosed by macrophages [tingible body macrophages (TBM)] or neighboring cells [9]. The main difference between the consequences of apoptosis and necrosis is that the apoptotic cell never leaks the cytoplasmic contents into the extracellular space and destroys the genetic content before phagocytosis. All these events are of great importance in the removal of cytotoxic T lymphocytes containing virus-infected cells and activated harmful granules. In this respect, apoptosis is a clean way of cell death [12].

Stress factors such as heat and cold shocks, oxidative stress, ultraviolet light, and ionizing radiation trigger the activation of stress-activated protein kinases (SAPKs) such as c-Jun N-terminal kinase (JNK) and p38. JNK is an important element of the signal pathway leading to apoptosis in response to stress conditions [9, 13]. Apoptosis related to our subject arises as cold-induced apoptosis, rather than the lethal effect of long-term storage, freezing damage or cold-struck damage occurs [14]. Contrary to the assumption that very low temperatures are lethal to cells, the cells are more severely damaged at moderate temperatures between −15 and−60°C. All chemical or biological reactions are almostly ceased at the liquid nitrogen temperature (−196°C). The only reaction that can occur at −196°C is the one arisen from cosmic radiation. A 200–400 rad dose of radioactivity can damage 63% of a cell population [15]. The earth is exposed to 0.1 rad cosmic radiation per year [16]. This corresponds to the amount of radiation that the frozen mammalian cell will be exposed to for 2000–4000 years at −196°C.

Cells that are stored at temperatures above −80°C, deteriorate over time and continue their cellular activities due to presence of solutes with different ion concentrations in the environment that are not fully frozen. Death can occur anytime, depending on temperature, animal breed, cell type, and freezing medium [16]. For a cell, which can be stored theoretically for 4000 years at −196°C, it can be considered that biological time has stopped [16].

4. Structure of biological membranes

All gamete cells, oocyte, and sperm, have a liquid mosaic membrane structure that are mainly composed of various phospholipids and proteins [17]. The cell membrane has at least three major tasks: to separate the cells from the external environment, to ensure that they have a specific shape, and to control the exchange of various solutes between the cell and the external environment.

The major structural components involved in the membrane structure are phospholipids, glycolipids, transmembrane proteins, peripheral proteins, and cholesterol. According to Parks et al. [18], membrane integrity is dependent on four important factors as follows:

1. Preservation of membrane integrity by bonds between the lipids to prevent lateral displacement of the lipids;

2. In order to preserve the stability of the lipids, they have to bond with the building blocks of the cell skeleton, with proteins forming a naturally occurring column;

3. The presence of large integral membrane proteins which serve to provide diffusion between the intracellular and the external environment and to stabilize intracellular balances; and

4. The presence of phospholipids and aquaporins with selective permeability in the membrane structure.

There are two layers, internal and external, formed by different types of phospholipids linked together in a chain form in the cell membrane. Positively charged phospholipids such as phosphatidylcholine and sphyngomyelin tend to be present in the outer leaflet of the cell membrane, whereas anionic phospholipids such as phosphatidylethanolamine, phosphatidylserine, and phosphatidylinositol tend to be present in the inner leaflet. The interactions between these two phospholipid layers work to form the transmembrane channels by cholesterol and proteins [18]. Phospholipids, which provide fluidic and variable membrane structure and are predominant in the docosahexaenoic acid chain, constitute 65–70% of total membrane lipids [19].

In order to make a more detailed explanation of the biological membrane structure, the first necessity is to identify the phospholipid and the above-mentioned sub-groups of phospholipids.

4.1 Phospholipids

Phospholipid consists of a phosphate group, one or two fatty acid groups attaching to this group, an alcohol group, and the glycerol or sphingosine backbone linking them. Phospholipids, which form the majority of the cell membrane, also increase the resistance of the cell to cold shock. It was also found that resistance to cold shock was higher in live sperm with high phospholipid/cholesterol ratio [20].

4.1.1 Phosphatidylcholine

It is a membrane lipid that consists of a phosphate group, two fatty acids, and the glycerol backbone of choline. Phosphatidylcholine, also known as lecithin, is present in high amount in the membrane and plays an active role on membrane stability. Simpson et al. [21] found that phosphatidylcholine exhibited protective properties against cold shock and that phosphatidylcholine-rich spermatozoa exhibited better motility levels than the spermatozoa deprived of phosphatidylcholine [21]. Sariozkan et al. obtained a higher level of motility in the group with 5% lecithine compared to the control group for bull sperm [22] (**Figure 1**).

4.1.2 Phosphatidylinositol

It is a membrane lipid that consists of a phosphate group, two fatty acids, and a glycerol backbone of inositol. It has functions in the growth and division of the cell, exchange of membranous materials, participating in the cytoskeleton structure and binding to the target proteins (membrane proteins) to maintain membrane stability and fulfilling their tasks [23]. Luconi et al. found that human sperm with phosphatidylinositol exhibited higher motility level [24] (**Figure 2**).

4.1.3 Phosphatidylserine

It is a membrane lipid consisting of two fatty acids and glycerol backbone of serine. It causes the start of capacitance or apoptotic changes through displacement between membrane foliage. Arrighi et al. [25] emphasized that phosphatidylserine is localized mainly in the head and middle region of a normal spermatozoon, emphasizing phosphatidylserine externalization in apoptotic spermatozoa, and 14% externally phosphatidylserine apoptotic spermatozoon in normal semen. Removing these externalized phosphatidylserine apoptotic spermatozoa from the environment, results in an increase in fertility. Wilhelm et al. [26] also investigated the effect of phosphatidylserine and cholesterol supplementation on sperm parameters after freezing/thawing, and consequently, samples with phosphatidylserine and cholesterol showed higher motility and viability than the control group (**Figure 3**).

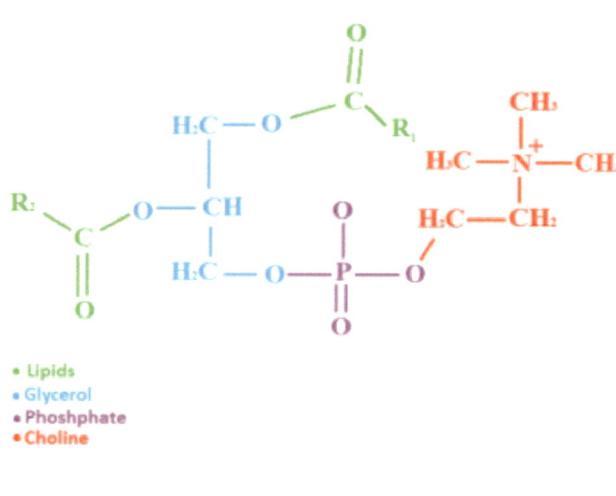

Figure 1.
Diagram of phosphatidylcholine molecule.

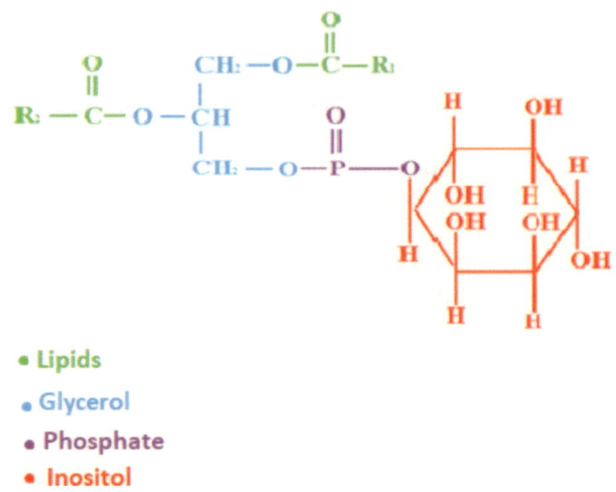

Figure 2.
Diagram of phosphatidylinositol molecule.

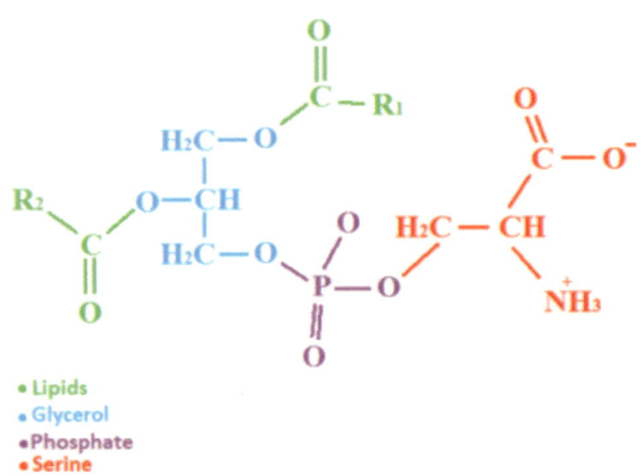

Figure 3.
Diagram of phosphatidylserine molecule.

4.1.4 Phosphatidylethanolamine

It is a class of lipids that consist of two fatty acids and a glycerol backbone of ethanolamine. It has the ability to build hydrogen bonds with other membrane proteins with its two electrons. This ability contributes to membrane integrity and plays a role in membrane reforming during phase transitions [27] (**Figure 4**).

4.1.5 Sphingomyelin

Unlike the other membrane lipids mentioned above, the backbone of sphingomyelin is formed by sphingosine (an unsaturated aminoalcohol) instead of glycerol. It is a membrane lipid consisting of a fatty acid and a sphingosine backbone of choline group (**Figure 5**).

Although phospholipids constitute the great majority of membrane structure, the most effective molecule is cholesterol that provides integrity to both the phospholipids and other components. The effect of cooling and freezing on the sperm

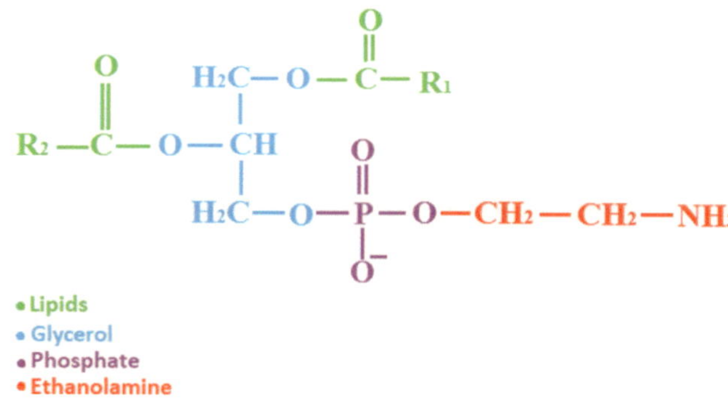

Figure 4.
Diagram of phosphatidylethanolamine molecule.

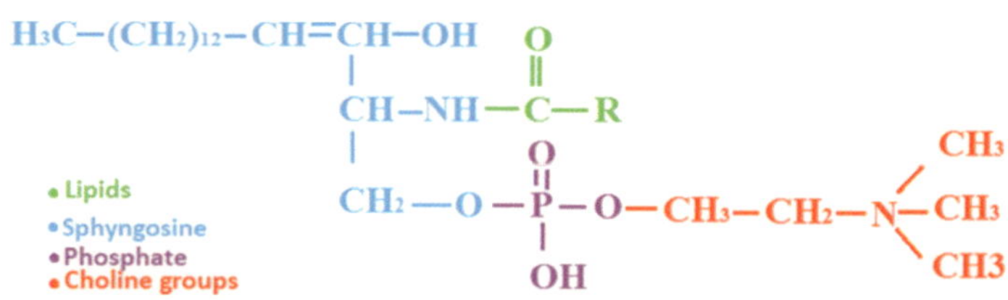

Figure 5.
Diagram of sphingomyelin molecule.

membrane is directly related to the cholesterol/phospholipid ratio in the membrane, saturation in the hydrocarbon chains (the chain formed by the addition of hydrogen bonds between the hydrogen and carbon atoms of the lipids themselves), and protein/phospholipid ratio [28]. At the same time, it was proven that cholesterol in phosphatidylcholine, phosphatidylethanolamine, and cholesterol mixtures was observed more intensely in the areas of phosphatidylcholine [29]. The cholesterol molecule is considerably smaller in size than phospholipids and therefore has the ability to move freely in the membrane. The cholesterol molecule is mostly concentrated in portions, where hydroxyl groups are rich in phospholipid ester carbonyls [30]. If cholesterol was extruded from the spermatozoon plasma membrane, there was an increase in fluidity, permeability, and fusion capacity in the membrane structure, and the structure of the membrane was damaged [31, 32]. When cholesterol was added into membrane, the membrane stability and motility increased, whereas capacitance decreased.

Cholesterol plays an effective role in the phase transitions in the membrane structure. As the cell membrane passes from the liquid crystal phase to the lamellar gel phase, the membrane cholesterol densifies the structure of the hydrocarbon chains of the fluid phospholipids and causes degradation of the lipids in the gel phase [33]. There are various proteins along with lipids on the cell membrane. Proteins, which have the functions of contributing to the membrane integrity by forming the attachment area to the lipids, and allowing the cells to exchange materials from the outside, are divided into two membrane proteins of, namely, the peripheral membrane and the integral membrane.

Peripheral membrane proteins can be found both on the exterior and interior of the cell. It has the task of forming protein-protein bonds and participating in the skeletal structure of the membrane. Integral proteins, which are another type of proteins, are opened both internally and externally of the cell and interact with both environments. It has functions, such as creating a holding place for cytoskeleton and glycoproteins, generating responses by receiving signals from various

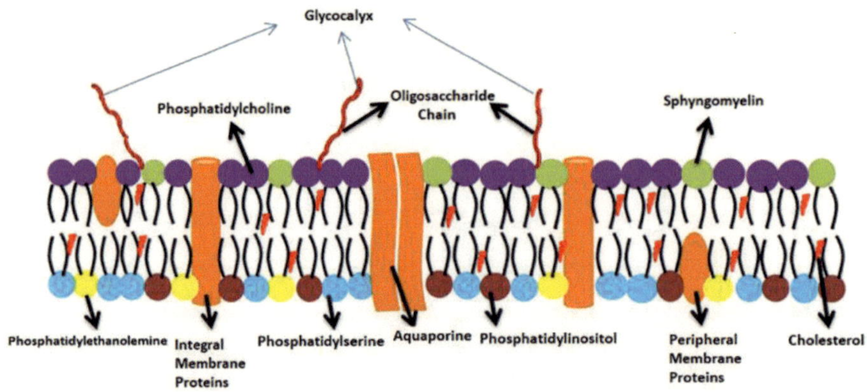

Figure 6.
A schematic representation of cell membrane components is shown.

chemical substances, and providing the exchange of substances between the inner and outer environments of the cell.

During freezing and thawing, protein channels called aquaporins (AQP) have functions to control the water inlet and outlet to the cell. Aquaporins are composed of four different monomers, which are independent of each other, in many cell types, from bacteria to plants and to mammalian cells. These four different monomers combine and form a tetramer. While each monomer is only water-permeable, the fifth channel in the middle allows passage of both water and various ions [34]. It is much faster and easier for water to pass through aquaporin channels when compared to membrane lipid. Although, there are 200 kinds of aquaporins in all living species, there are only 13 kinds of aquaporins in mammals. AQP7, AQP8, and AQP11 were identified on the spermatozoon membrane. While AQP7 is responsible for sperm glycerol metabolism, AQP8 is responsible for the water exchange of the cell. AQP11 is involved not only in the membrane structure, but also in the tail as well as in the exchange of cellular matter [35].

Cells use oligosaccharide chains (sugar chains formed by combining 2–20 monosaccharides) that extend out of the membrane as a tentacle or an arm to perceive various structures in the extracellular environment, interact with other cells, and initiate the necessary physicochemical reactions. These sugar chains are termed glycolipid chains when derived from lipids and glycoprotein chains when derived from proteins. These glycolipids and glycoprotein chains extend outward to bring a network-like structure around the cells. This network-like structure is called glycocalyx (**Figure 6**).

4.2 Damage and phase changes in the membrane structure

The cell membrane is a fluidic mosaic structure with cholesterol, integral and peripheral proteins, lipids, and many more.

There are many different types of phospholipids in the cell membrane. The changes that occur during freezing-thawing also originate from the structural differences in these phospholipids. The main factor that drives the structural differences is the dimensional changes between the head part forming the phospholipids and the acyl chains. These dimensional changes directly affect the various changes in the membrane structure at the later stage as well as lipids depending on the lipid type.

If in a phospholipid, the head and acyl chains are similar or identical in size, the phospholipid molecule is in a cylindrical form. Phosphatidylcholine is as an example for this kind of phospholipids. This kind of phospholipids is termed as bilayer lipids and are not significantly affected by phase changes during cell cooling. In the homogeneous phospholipid distribution that occurs during the phase change, these types of phospholipids in the form of bilayers are gathered together to form the bilayer membrane model; in other words, they continue to create the lamellar phase of the liquid crystal. If in a phospholipid the head and acyl chain are in different sizes, the phospholipid is in an inverted or flat cone-like form. This kind of phospholipids is called non-bilayer form phospholipids. Phosphatidylethanolamine is as an example for this type of lipids.

When the cell starts to cool down, the non-bilayer form of phospholipids undergoes a phase transition from the liquid crystal phase to the gel phase [36]. If the head and acyl chains are of different sizes in a phospholipid, this kind of lipids are also called non-bilayer lipids. During the phase changes, the non-bilayer lipids in the aggregate come together to bring up the different structures, which will be mentioned in the next section. Non-bilayer form phospholipids mainly consist of 2 types of phospholipids:

Type 1: Polar (+charged, hydrophilic) head in type 1 non-bilayer phospholipids is larger than apolar (−charged, nonpolar) tail. For this reason, they are in the form of an inverted cone. When the membrane is cooled, the negatively charged heads are directed outward and the neutral apolar parts are inwardly arranged to form micelle-like structures depending on temperature changes [37].

Type 2: For polarity of type 2 form phospholipids, the polar head is smaller than the apolar tail. For this reason, they are in the form of a flat cone, and when temperature changes, the negatively charged head part is arranged in the apolar neutral part outwardly and called hexagonal II (H_{II}) [38] (**Figure 7**).

Non-bilayer lipids are distributed heterogeneously on the cell membrane and carry out some vital tasks for the cell, including:

1. providing membrane flexibility,

2. participating in the barrier activity against the outer environment,

3. protecting their stability by binding to the membrane's peripheral proteins through creating a surface to attach to the proteins, and

4. building the structures that integral proteins can hold, etc.

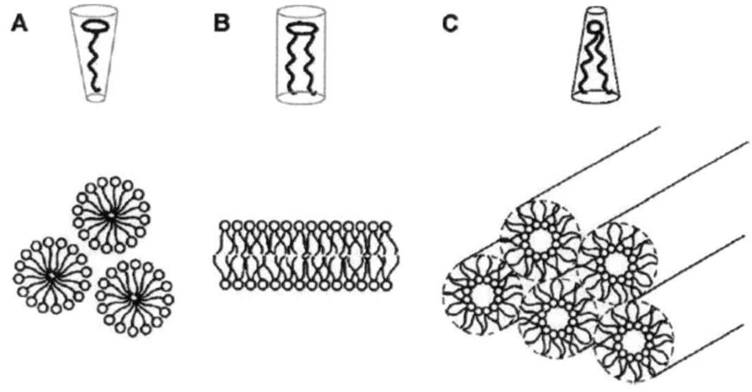

Figure 7.
This schematic representation depicts phospholipid-shaped non-bilayer type 1 lipids as indicated by the letter A and they are schematized at the bottom just below. In B, bilayer form phospholipid is schematized and bifilar lamellar form with a normally arranged phospholipid model was also depicted in the case where the cells were not subjected to heating or cooling (at a 36°C temperature). In C, non-bilayer type 2 lipids and hexagonal form II formulated due to heat exchange are schematized [37].

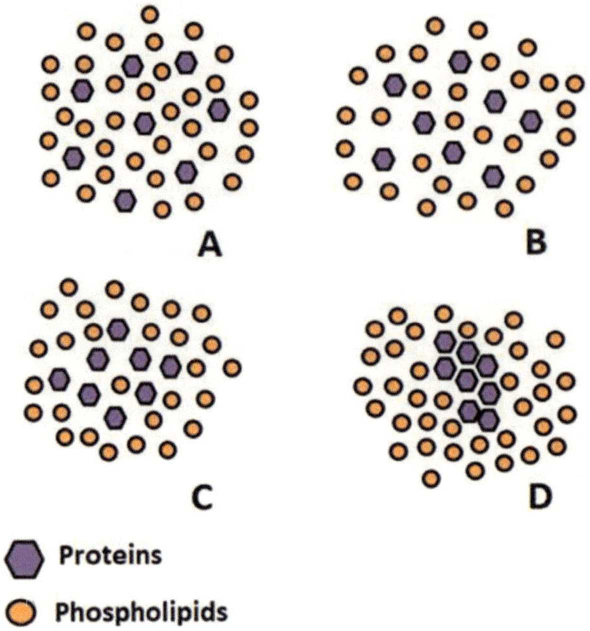

Figure 8.
In this schematic, the effect of cooling on the cell membrane structures is shown. The phospholipid-protein components, which are in a heterogeneous structure starting from A, show clusterings due to cooling (B, C) as mentioned above, and membrane proteins appear to be tight in the center while phospholipids appear to be like honey pellets in the last stage (D).

Membrane phospholipids enter a phase transition generally between 36 and 5°C. These phase changes are species-specific and cause different responses to temperature changes depending on the animal's spermatozoa. When cell cooling starts, some phospholipids in the membrane structure undergo phase changes. The phospholipids in the non-bilayer form are bound to each other by hydrocarbon-chain bonds. As the hydrocarbon chains undergo crystallization and hardening, the lipids passing through the hexagonal Form II become more tightly bonded to each other, and the membrane proteins weakly bound to phospholipids cause aggregation on the membrane. They are tightly agglomerated to occupy less space than normal (**Figure 8**).

The phospholipids that are interlocked with the proteins and hydrocarbon bonds that come together due to phase changes during cooling of the cells are in the nonlamellar phase at −196°C. However, during thawing, proteins and phospholipids do not return to their original situations. In this process, the membrane in the form of non-bilayer phospholipids came together and converted into the structures in the form of phospholipid micelles and hexagonal Form II. However, during thawing, these phospholipids cannot recombine to form bilayer phospholipids and turn into a fluidic mosaic membrane structure; therefore, the membrane structure has permanently been damaged (**Figure 9**). In this case, the non-bilayer phospholipids interacting with proteins are separated from the proteins and converted into a homogeneous form, different from the mosaic form.

Since proteins and various phospholipids are arranged in a heterogeneous structure in the normal arrangement of cells, exchange of substances is achieved in almost every region of the cell membrane. When the membrane structure is deteriorated in this way, substance exchange occurs especially where protein aggregation is intense. In this case, the non-bilayer phospholipids in specific interactions with proteins are separated from proteins and transformed from fluidic mosaic to homogeneous form [40].

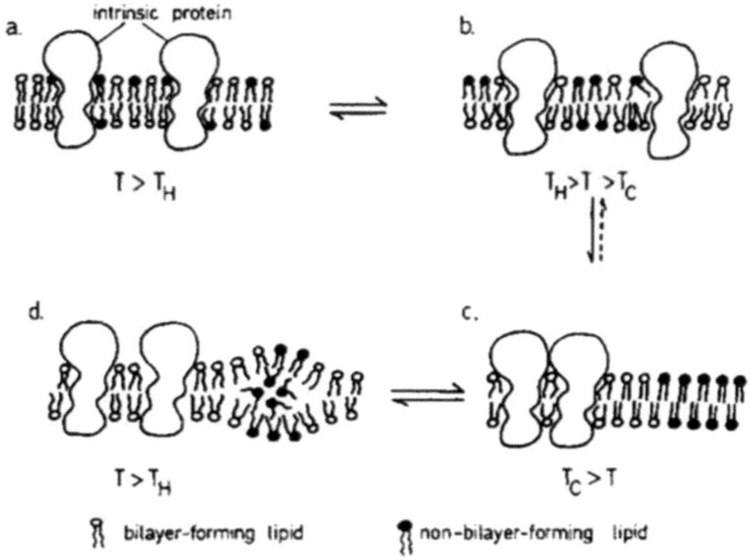

Figure 9.
Bilayer and non-bilayer phospholipid interactions during cooling and after re-thaw process [39]. The figure shows the phase separations in biological membranes associated with cooling biological membranes from their growth temperature (a) to a temperature below the gel (b) to liquid-crystalline phase-transition temperature (c) and reheating to the growth temperature (d). Lipids that tend to form hexagonal-II phase are indicated by a solid head group.

Proteins are one of the basic building blocks of the cell and carry out many functions in the cell metabolism. Hydrogen bonds in the water molecules play a very important role in the denaturation of proteins. In the hydrophobic polar (HP) protein model, amino acids are reduced to polar (represented by P, charge or dipole) or nonpolar (H) point entities. Hydrophobicity is defined as the desire to reduce surface area of proteins that are interacting with water, and is related to the conversion of neighboring molecules into a similar energy-attractive structure. The hydrophobic molecules in the solution produce a more energetically favorable structure compared to the water bodies by forming ice-like cages in their surroundings at low temperatures. The cell systematically tries to reduce the energy by increasing the number of cages. This is the cold denaturation of proteins [41]. On the opposite side, at higher temperatures, water molecules do not form cages. Irregular water molecules around the hydrophobic amino acids in proteins are energetically less advantageous than the water bodies. Thus, the proteins try to hide their hydrophobic parts in the building blocks. Biological structures, especially proteins, undergo structural changes during temperature changes. For this reason, nucleic acids and many polysaccharides need to stay in a certain temperature range with specific ionic strength and pH value to fulfill their functions. Cold protein denaturation occurs at a temperature range of 0 to −20°C, reversibly or irreversibly, but mostly irreversibly. Globular proteins often undergo partially reversible denaturation. During denaturation, the covalent bonds between the atoms forming the peptide bonds between the amino acids of the proteins are not broken, but only hydrogen bonds change their structure. Protein denaturation during freezing or thawing leads to protease leaks from liposomes presenting in the acrosome and loss of membrane integrity.

5. Cryoprotective agents (CPA)

Prior to the discovery of cryoprotective agents, some successful cooling protocols relied on extracellular ice formation suppression and sustaining cell viability by various chemical agents [7]. Mammalian cells had not been successfully cryopreserved until 1949 by Polge, Smith, and Parkes when glycerol was accidentally found to have protective functions in freezing cells [3]. It has also been observed that if the correct dose of cryoprotectant is used and the cells are cooled down at the optimal cooling rate, the survival potency of cells is increased [42].

Cryoprotectants are used to avoid or decrease the cold shock damage and intracellular ice formation during freezing, recrystallization during thawing, and membrane destabilization. Cryoprotectants can reduce the freezing point and the proportion of salts and solutes in the sample by increasing the amount of liquid fraction, and suppress ice formation both outside and inside of the cells [43]. In biological structures, while the hydrogen bonds between the membrane phospholipids are connected by the oxygen atoms contained in the water molecules, the cryoprotecting substances such as glycerol are replaced by water. These bonds have utmost importance for membrane integrity [44]. It has been reported that supplementation of 0.2 M sucrose or trehalose into bull semen during freezing and thawing increases sperm viability [45]. A good cryoprotectant should be water-soluble and have minimal toxic effects [46].

Cryoprotectants can be divided into two groups according to their mechanism of action, as penetrating and non-penetrating cryoprotectants.

5.1 Penetrating cryoprotectants

Penetrating cryoprotectants have low molecular weight and therefore they have the ability to enter the cells. In this way, they can affect both intracellular and extracellular environments.

When cryoprotectant is added into the cell suspension, cytosolic water moves to the exterior milieu of the cell due to the water chemical potential difference between inside and outside of the cells. The penetrating cryoprotectant penetrates into cells because of the concentration difference. This process lasts until an equilibrium between intracellular and extracellular environments for both water and cryoprotectant is reached. The freezing point of the intracellular medium is decreased, and the intracellular ice formation (IIF) can be eliminated or prevented. Penetrating cryoprotectants penetrate into the cells, form new hydrogen bonds with water molecules by breaking the hydrogen bonds between them, and thereby change the structure of water. In this way, the cryoprotective function exhibits by preventing the cells from reaching high concentration of ions and avoiding extreme dehydration due to water loss during freezing [47].

Penetrating cryoprotectants such as glycerol, dimethyl sulfoxide (Me_2SO or DMSO), ethylene glycol, formamide, 1,2-propanediol, 2,3-butanediol, and propylene glycol affect spermatozoa during cryopreservation by changing the physical properties of intracellular solution, decreasing the intracellular ice formation, increasing the resistance to cold shock, regulating the protein and lipid structure of the cell membrane, and increasing membrane fluidity[48]. In the following part, two most widely used penetrating cryoprotectants, glycerol and DMSO, will be discussed.

5.1.1 Glycerol

Many penetrating cryoprotectants have been tested, giving different results in cryopreservation of gamet cells of different species. However, the most commonly used penetrating cryoprotectant is still glycerol, which was discovered in 1949 by Polge et al. [3] as a result of sperm vitrification.

As shown in **Figure 10**, a glycerol molecule contains three hydroxyl groups. Each glycerol molecule is capable of binding to three water molecules. Because of the

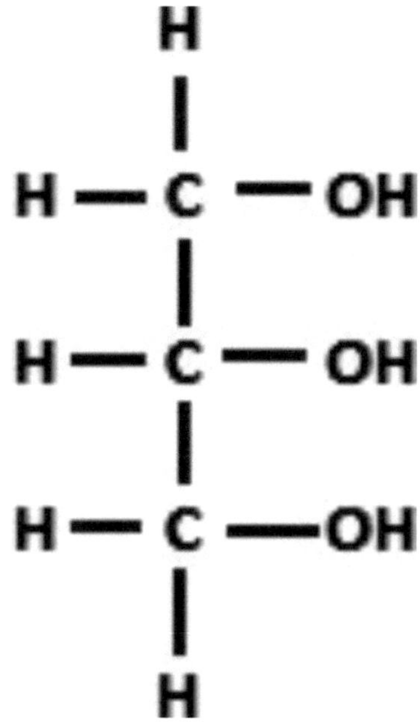

Figure 10.
Diagram of a glycerol molecule.

smaller molecular size, glycerol can easily pass through the membrane pores. With the decrease of temperature, the hydroxyl bonds formed between water molecules get hardened and solidified. For this reason, the frozen water expands and damages the cell. As a cryoprotectant, glycerol exerts its function by preventing water molecules from becoming large-volume ice crystals by binding to the hydrogen atom of the water molecules and interlocking during freezing.

In addition to the osmotic effect on cells, glycerol directly acts on the cell membranes. Amann and Picket [49] have shown that glycerol exerts its primary effect in the extracellular environment, but it also intracellularly affects the cell membranes and organelle membranes. It binds to phospholipid groups in the plasma membrane, reduces the membrane fluidity, [50] and also forms a particulate clump in the membrane by interacting with membrane proteins and glycoproteins [51, 52]. At the same time, it creates structures similar to gap junctions (small connection channels between cells) in the membrane, reduces the electrical capacitance of the membrane, and causes large-scale rearrangement of membrane structure [53]. Glycerol causes a decrease in the membrane fluidity by providing interactions between the inner and outer layers of the membrane through regulating the structure of fatty acid acyl chains. Glycerol also affects the polymerization (creating multiple monomers to form larger and differently shaped macromolecules, called polymers) and depolymerization (decomposition of large macromolecules into smaller monomers) of microtubules, which indirectly affect the plasma membrane. Changes in the microtubule structures in the tail of spermatozoon influence the interactions between microtubule-associated proteins [54].

While glycerol has many benefits as a cryoprotectant, its use in high dose has a detrimental effect on the cells. These harmful effects appear in at least seven cellular activities:

- It denatures proteins [46].
- It increases the viscosity of cytosol.

- It causes changes in the polymerization and depolymerization of α and β tubulins, the basic proteins of the microtubules found in the spermatozoon tail.
- It causes structural changes in the microtubuli.
- It acts on the bioenergetic balances.
- It acts directly on the plasma membrane and glycocalyx (a meshwork cell coat that is formed mainly by external proteins, externally-located glycoproteins and glycolipids providing various chemical interactions between the cell and the external milieu) [55].
- It creates osmotic stress due to slow penetration through the cell membrane than other cryoprotectants [56].

Si et al. measured the motility and integrity of both sperm and acrosomal membranes that were frozen with different concentrations of glycerol (2, 5, 10, and 15%) and dimethyl sulfoxide (2, 5, 10, and 15%) in their study on Rhesus monkeys. The highest motility was measured as 45.5% in the 5% glycerol group. In terms of membrane integrity, the control group had the highest score (77.9%), while the 5% glycerol group had lower value (61.6%). Values of acrosomal integrity were 91.2 and 82.4% for the control group and 5% glycerol group, respectively. When these two parameters taken into consideration, DMSO had lower augmenting effects on Rhesus monkey semen than glycerol [57].

Awad and colleagues compared glycerol, ethylene glycol, and methanol as cryoprotectants and measured the sperm motility post thawing. The best result was obtained with 3% glycerol (72.4% motility). Motilities of sperm cryopreserved with 3% ethylene glycol or 3% methanol were significantly lower (56.9 and 22.6%, respectively) [58].

5.1.2 Dimethyl sulfoxide (DMSO)

Dimethyl sulfoxide (DMSO) is a liquid at room temperature with color scale changing from colorless to yellow. It is soluble in water, ethanol, acetone, diethyl ether, benzene, chloroform, as well as in aromatic compounds. Apart from being a cryoprotectant for semen, DMSO also has anti-inflammatory properties, bacteriostatic and tranquilizing effects, diuretic and local analgesic activity. It is a penetrant and carrier, and it can strengthen the effect of insulin [59] (**Figure 11**).

When used in concentration of 5–10%, DMSO protects against damage on mouse T and B lymphocytes [60], human embryos [61], and many cellular systems during freezing and thawing. Sometimes, DMSO has even higher protective potential than that of glycerol in freezing sperm of some animal species, such as elephants [62, 63]. Despite that DMSO has the same mechanism of action with glycerol, DMSO has lower penetrability in human and porcine spermatozoons [64]. At the same time, DMSO has better protective capacity than glycerol and ethylene glycol in sperm freezing/thawing process and is more active in testicular tissues than glycerol and 1,2-propanediol [7].

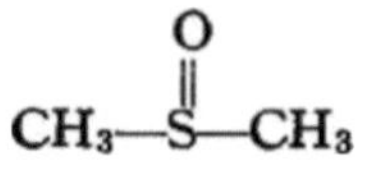

Figure 11.
Diagram of DMSO molecule.

Anchordoguy et al. used different cryoprotectants (DMSO, glycerol, proline, sucrose, and trehalose) for shrimp sperm and measured the viability. DMSO showed the highest viability (56.3%), followed by glycerol (29.3%). Proline and sucrose gave similar viability (24.5%), whereas trehalose gave relatively lower viability (16.2%). The group without cryoprotectant did not show any viability after thawing [65].

Two used different concentrations of DMSO and glycerol in this study of cryopreservation of semen of black grouper fish and graded the motility scores from 0 to 4. The level 0 means no motility, 1 means 0–25% motility, 2 means 25–50% motility, 3 means 50–75% motility, and 4 means 75–100% motility. Results of the study revealed that 10% DMSO resulted in level 3 of motility, 20% DMSO resulted in level 4, and 30% DMSO resulted in level 1. However, glycerol solutions with different concentrations showed level 0 motility [66].

5.2 Non-penetrating cryoprotectants

Non-penetrating cryoprotectants cannot penetrate through the sperm membrane and only act in the extracellular environment. They also function by incorporating the membrane structure or reducing the freezing point of the medium [48]. They are generally divided into two groups: low molecular weight cryoprotectants and high molecular weight cryoprotectants [45]. Non-penetrating cryoprotectants with low molecular weight include three subgroups as monosaccharides (glucose and galactose), disaccharides (sucrose and trehalose), and trisaccharides (raffinose and melezitose). Cryoprotectants with high molecular weight [e.g., polyethylene glycol (PEG), ficoll 70, polyethylene oxide (PEO), polyvinyl alcohol (PVA), polyvinyl pyrrolidone (PVPP), etc.] exert their functions by reducing the ice crystals formed externally during the freezing/thawing process to decrease cellular damage. As a special and effective cryoprotectant, trehalose will be discussed as follows.

Sugars used during sperm cryopreservation serve as an energy source and an osmotic pressure regulator, cryoprotectant, and cellular ice formation reducer [67]. As a disaccharide formed by binding two D-glucose molecules, trehalose extends the distance between the membrane phospholipids by binding to the polar part of the phospholipids in the cell membrane. These cavities formed in the membrane prevent the formation of ice in the cell by helping the outflow of water from the cell during freezing, preventing harmful effect of cellular dehydration, and stabilizing the cell membrane.

Phospholipids accumulate due to van der Waals forces when energy is drawn from the environment by cooling sperm and transition occurs from liquid crystal phase to gel phase. During thawing, irregular voids occur in the cell membrane. This causes damage in the membrane structure, irregular ion and water leakage both into and out of the cell [68]. When trehalose is added into the medium, it forms gaps between phospholipids in the cell membrane and these gaps prevent the phospholipids from aggregation during freezing and protect the entity of membrane structure after thawing [69] (**Figure 12**).

In a study by Bucak et al. on freezing ram semen, two different doses of glutathione (5, 10 mM), taurine (50, 100 mM), and trehalose (50 and 100 mM) were used as cryoprotectants, and motility of the semen samples were measured at 0, 6, 24 and 30 h post thawing. The highest motility was obtained in samples containing 50 mM trehalose [67]. Hu et al. investigated cryopreservation capacity of trehalose on five different groups of bull sperm. Group 1 was the control group without trehalose, whereas groups 2–5 contained trehalose with concentrations of 25, 50, 100 mM trehalose, and 200 mM. The group with 100 mM trehalose had the highest semen motility, mitochondrial activity, acrosome integrity than those of the other

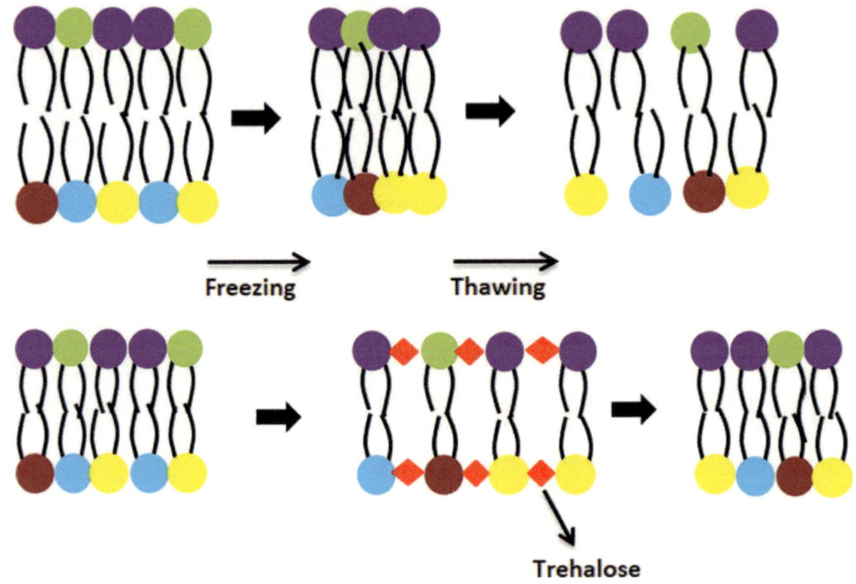

Figure 12.
Schematic drawing of cryoprotective effect of trehalose on membrane phospholipids.

groups [70]. Öztürk et al. studied the effect of arginine and trehalose on bovine sperm samples in their study. Trehalose showed higher (50.5% subjective, 64.7% CASA) motility than arginine (29% subjective, 12.2% CASA) when compared to the control group [71]. Uysal and Bucak studied rapid (10°C/min from 5 to −25°C, 50°C/min from −25 to −130°C) and slow (0.5°C/min from 5 to −25°C, 50°C/min from −25 to −130°C) cooling rates with different doses of trehalose (0, 50, 100, 150 mM), and investigated sperm motility, survivability, membrane integrity, and morphologically abnormal sperm rates after thawing. Evaluations showed that the medium with 100 mM trehalose resulted in the best scores (sperm motility 72.0%, morphologic sperm abnormalities 25.5%, sperm viability 75.9%, and sperm membrane integrity 68.2%) [72]. Bucak and his colleagues also used trehalose (50, 100 mM), taurine (25, 50 mM), cysteamine (5, 10 mM), and hyaluronan (0.5, 1 mM) as antioxidants in their study of ram sperm cryopreservation. Motility, acrosomal damage, viability, and total abnormality parameters were examined. The group with 50 mM trehalose, 25 mM taurine, and 5 or 10 mM cysteamine provided the best results [73].

6. Molecular structure of water in liquid and solid phases

As the main source of life, water is different from most of the other molecules in the nature. Depending on the amount of energy in the environment, the water molecules undergo changes in their structure and phase transitions occurring between liquid, solid, and gas phases. Water molecules are in the liquid phase at the mammalian body temperature and perform very important tasks in almost every living cell.

The water molecule consists of two hydrogen atoms binding to a negatively charged oxygen atom at an angle of 104.52° with covalent bonds of 95.84 picometer in length [74] (**Figure 13**). Hydrogen bonds are formed between water molecules (**Figure 14**). The fluid structure of the water at room temperature is based on the fact that these bonds are constantly breaking and re-bonding [75].

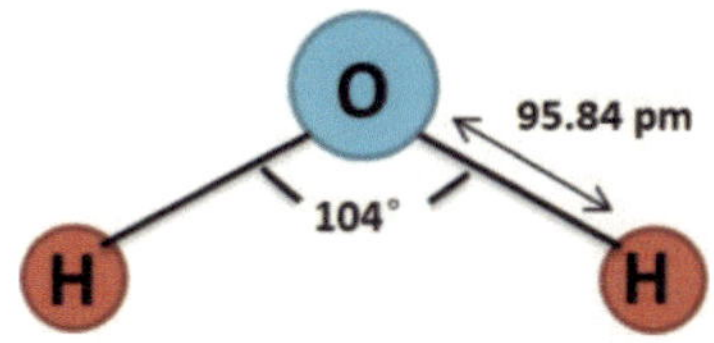

Figure 13.
Schematic drawing of the hydrogen bridge bond.

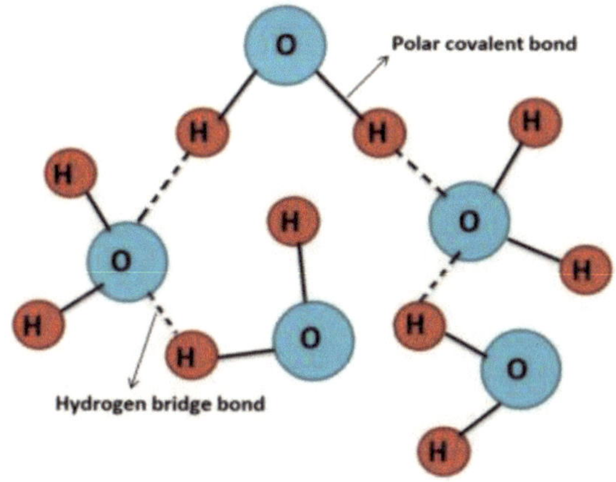

Figure 14.
Schematic drawing of water molecules.

As mentioned earlier, water molecules are in liquid form in living organisms. However, during cooling, the covalent bonds between oxygen and hydrogen are shortened and the hydrogen bonds become hardened and extended. So, they tend to line up in a symmetrical manner and to expand. Important effects on water-based biological molecules should not be overlooked. The water itself is described as a biomolecule, even as the 21st amino acid [76]. In some cases, the ability of protein molecules to perform their normal functions depends on the presence of water molecules on their surface [77]. The bond between water and protein surface is formed by giving the proton in the hydrogen atom of water to the protein or by taking the proton in the hydrogen of the amine group and its derivatives (NH, NH_2, NH_3), which forms the backbone of proteins. Water molecules also build bridges ("water bridges") to provide interactions between the atoms of different protein molecules [78].

7. Cryobiological events during freezing and thawing

During cooling and thawing, spermatozoa undergo various changes in their chemical and physical status. The first one is the phase change of lipids in the sperm membrane, generally happening between 17 and 36°C. Different lipids have different phase transition temperatures. Proteins are normally found in asymmetric position between the membrane lipids in the liquid crystal structure. When the lipids transform to gel phase, membrane structures are disrupted and ion metabolism is regulated. When the membrane lipids are in liquid crystal or gel phases, the leakage of liquid electrolyte through the membrane is minimal. However, during phase changes of lipids, the member permeability is increased [79], which causes solute leakage. Lipid phase changes occur not only during cooling but also during thawing. Fluorescent diacetate leaching was observed in the spermatozoa treated with fluorescent diacetate and re-dissolved by cooling to −5, −10 and−20°C [19].

The second change that takes place is the transformation of the cytosolic water in the cell into ice. Spontaneous ice formation occurs between −5 and−15°C. Generally, above −5°C, the internal and external milieu of the cell are supercooled, and no ice formation is observed. At the temperature between −5 and−10°C, ice formation starts in the extracellular environment, then cells are dehydrated and intracellular environment is supercooled. The fate of the cryopreserved cells, life or death, depends on the following cooling and thawing process [28].

7.1 Cryobiological damages during freezing

As mentioned in the two factor hypothesis proposed by Peter Mazur, damage arising from freezing occurs through two components. The first one can be defined as the direct damage caused by ice (ice injury), and the second one as the effect of the solution ion concentration on the cell (solution injury) [2].

However, this situation appears in different manners in male and female gametes. While intracellular ice formation is observed in mammalian oocytes and embryos at different degrees of cooling [80], the case is not the same for spermatozoa [81]. The main factors include the pressure on spermatozoa caused by ice crystals formed in the external environment and the harmful effect of the high concentration of salt and mineral inside the cell.

7.2 Two-factor hypothesis of cryoinjury

The two-factor hypothesis suggested by Peter Mazur can be explained as follows:

Factor 1: During cooling, extracellular ice formation happens first, which leads to the increase of solute concentration outside of cells. If the cooling rate is slow enough, the cells will have enough time to dehydrate and the cytosolic water will outflow from the cells until it is balanced with extracellular water. Dehydration of cells may include a few steps: separation of water molecules from the cytoplasmic liquid environment to the lipid part, diffusion in the lipid bilayer, and migration from lipid to external fluid, all of which are associated with activation energy (threshold energy value for a chemical reaction to takes place) (**Figure 15**). In this case, injury to the cells is majorly from the highly concentrated solutes in the cells [16]. Some biological constructs (e.g., eight-cell embryos) may be resistant to solution effects [82].

Factor 2: If the cooling rate is rather high, the cells will not have enough time to discharge cytosolic water into the extracellular environment, and water will freeze inside the cells [16] (**Figure 16**).

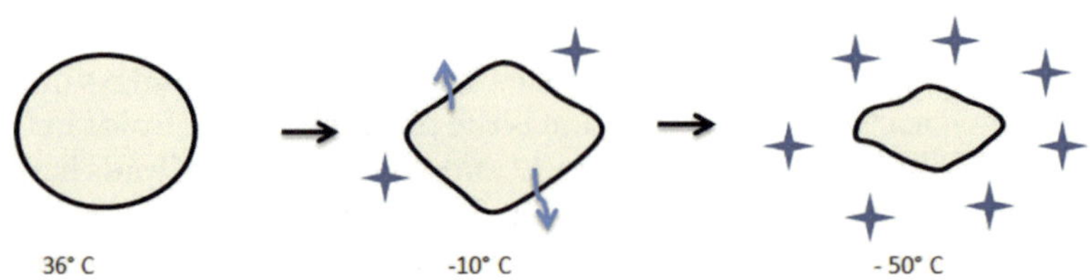

Figure 15.
Schematic of factor 1 in "two-factor" hypothesis of cryoinjury. During slow cooling procedure, intracellular water has enough time to outflow from the cells.

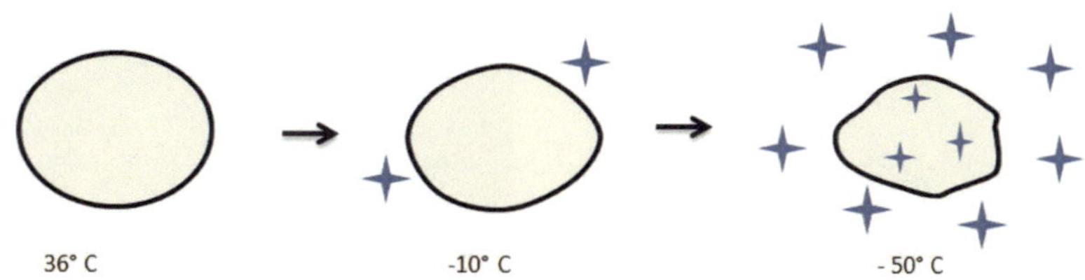

Figure 16.
Schematic of factor 2 in "two-factor" hypothesis of cryoinjury. If cells are cooled rapidly, water does not have time to outflow and freezes both inside and outside of the cells.

In more detail, when the cells are cooled slow enough, the cells dehydrate and an increase in the intracellular solute concentration is observed before reaching the freezing temperature. In long term, the cells are damaged due to severe dehydration and the increased intracellular ion concentration, which allows the exchange of some critical ions (e.g., potassium) via channels in the membranes [83]. The increase in the ion concentration causes the water-insoluble proteins in the cell to interfere with these ions and transit into a soluble form. The osmotic pressure remains unchanged as these ions bound to proteins causing an artificial decrease in intracellular ion concentration. Protein concealment of intracellular potassium and chloride ions allows the entry of sodium from exterior, and the cytosolic ion concentration increases again. During defrosting, the extracellular ice first melts, dilutes the outer environment, and water enters the cells. As the amount of cytosolic water increases, the potassium and chlorine-binding proteins liberate from the bonds and the hidden potassium and chloride reappear. As the amount of intracellular ions increases, the cells begin to rehydrate, during which process cracks in the cell membrane and cell lysis were observed [82].

On the other side, if the cells are cooled down too quickly, the intracellular water will not have enough time to balance with the extracellular environment, freeze inside the cells, and damage the intracellular structure and cell membrane [6]. The inhibition of intracellular ice formation by cryoprotectants is undoubtedly helpful on the survival of the cell [84]. However, the increase in the amount of intracellular solution leads to damages such as protein denaturation, cell lysis (cell membrane breakage), and mitochondrial and nucleus damage [85].

The morphological and physiological differences in the spermatozoa of different animal species must be taken into consideration. Thus, the sperm of different species will have different responses to cryopreservation; therefore, the optimal cryopreservation protocol would be different for them. For example, the optimal cooling rate was 100°C/min for the bull sperm, whereas 1–10°C/min for human [84], 30–50°C/min for pig [85], and 50–60°C/min for ram [45, 86].

Morris used different cryoprotecting agents (CPAs) in the sperm cryopreservation with both rapid and slow cooling procedures (**Figures 17** and **18**) [87]. Bucak et al. studied the protective effects of different antioxidants added to the diluent medium on cryopreserved spermatozoa (supported by the Scientific and Technological Research Council of Turkey, project No: 114O642). In this study, sperm cryodamages were ultrastructurally demonstrated. In the Scanning Electron Microscopy (SEM), vacuole-like structure and head, neck, tail deformations were visualized (**Figure 19**) [88]. In the Scanning Transmission Electron Microscopy (STEM), membrane deformations, head, axoneme, acrosome, mitochondrium damages, and double tail were observed (**Figure 20**) [88].

Ice crystals formed in the extracellular medium also act on the cytosol and may result in intracellular ice formation. When cells are in close proximity or in contact

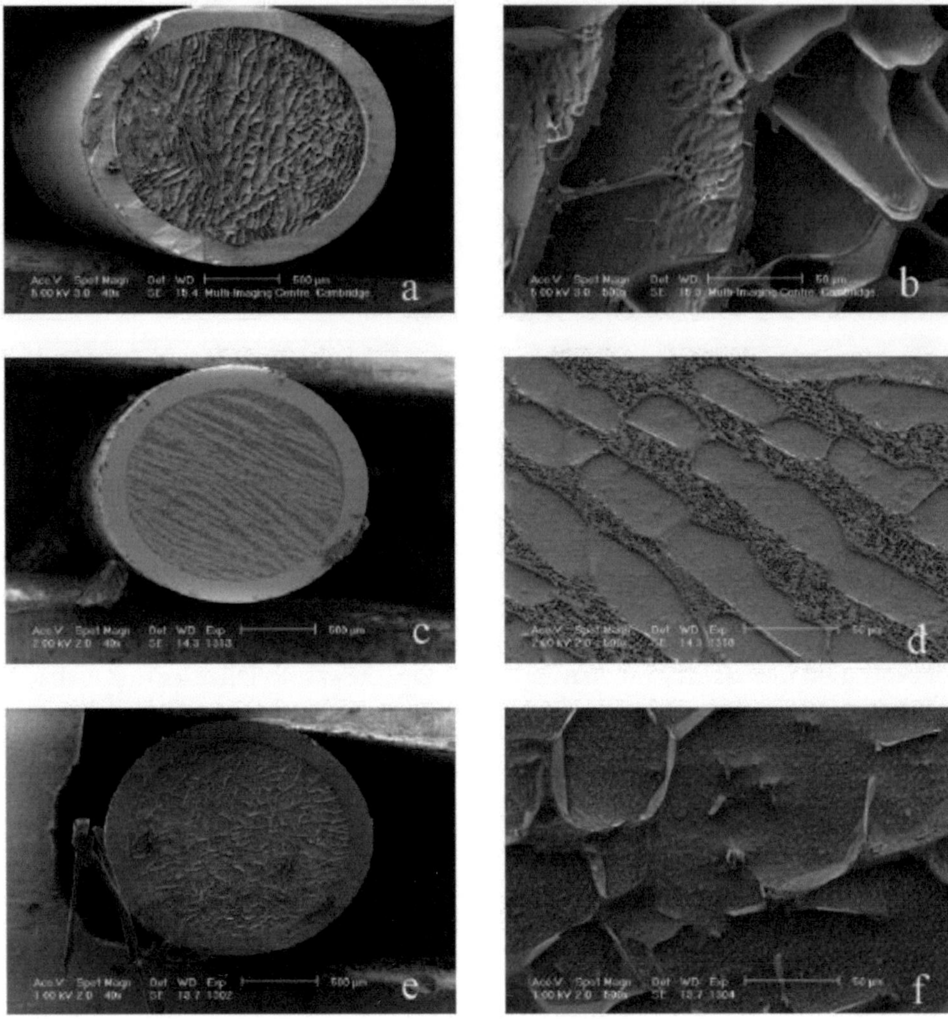

Figure 17.
In (a) and (b), samples with added glycerol (3 volumes of 0.1 M sodium citrate to 1 volume of 0.33 M fructose and 1 volume of 0.33 M glucose + 4 ml of egg yolk +3 ml of glycerol) were frozen at 10°C/min visualized with a Cryo-scanning electron microscope. In (c) and (d): Glycerol added samples were frozen at 3000°C/min. In (e) and (f), sperm were frozen at 3000°C/min without glycerol. In (a), there are channels occurs (in which the spermatozoa can be protected) under the effect of glycerol, whereas (c) shows narrow channels as opposed to (a). In (e) no cryoprotectant substance is added and it is observed that the formed ice does not leave enough space for spermatozoa. This is illustrated in more detail in **Figure 18** *[87].*

with any of extracellular ice crystals, the aquaporin channels in the cell membrane are also affected. Small ice crystals penetrate through the aquaporins into the cytosolic milieu and may trigger intracellular ice formation [34].

7.3 Damage during thawing

Rapid cooling causes the formation of small intracellular ice crystals. Such small ice crystals have higher energy than large ice crystals because of their small radius curves and they are thermodynamically more irregular. Due to this free energy difference, they show melting at lower temperatures than large ice crystals. Recrystallization is a phenomenon that small ice crystals combine together into larger ones, which cause defects such as various cuts and defects on organelles and membranes in the cell by physical pressure or crystals [87, 90]. Damage due to

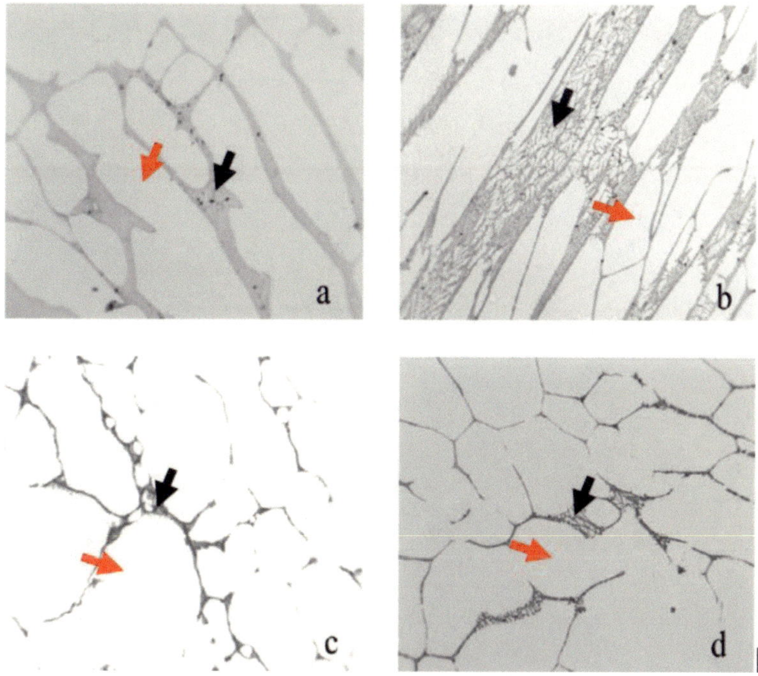

Figure 18.
In the above photographs, glycerol was added and frozen at a rate of 10°C/min (a), glycerol added and frozen at 3000°C/min (b), frozen at 10°C/min without glycerol (c), and frozen at 3000°C/min without glycerol (d) samples are illustrated. In the photo shown with the letter A, there is an unfrozen section between the ice masses (red arrow), where cells locate themselves, in the photograph b, the same frozen part is observed but in this part there are small pieces of ice in pieces and it poses a danger to the cells. While there are much smaller frozen canals on c photo, there is small ice crystals in the d picture, even in the unfrozen canals (black arrow) [87].

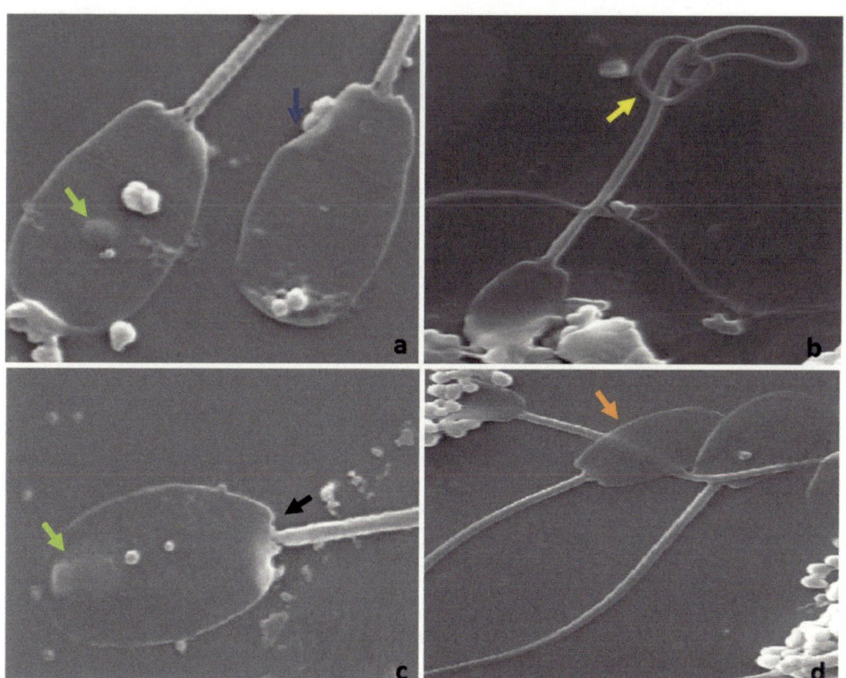

Figure 19.
SEM micrographs of frozen-thawed spermatozoa. (a) Vacuole-like structures (green arrow), head damage (blue arrow). (b) Tail damage (yellow arrow). (c) Vacuole-like structures (green arrow), neck damage (black arrow). (d) Normal sperm (orange arrow). a, c: 10,000×, b: 5000×, d: 3000× [88].

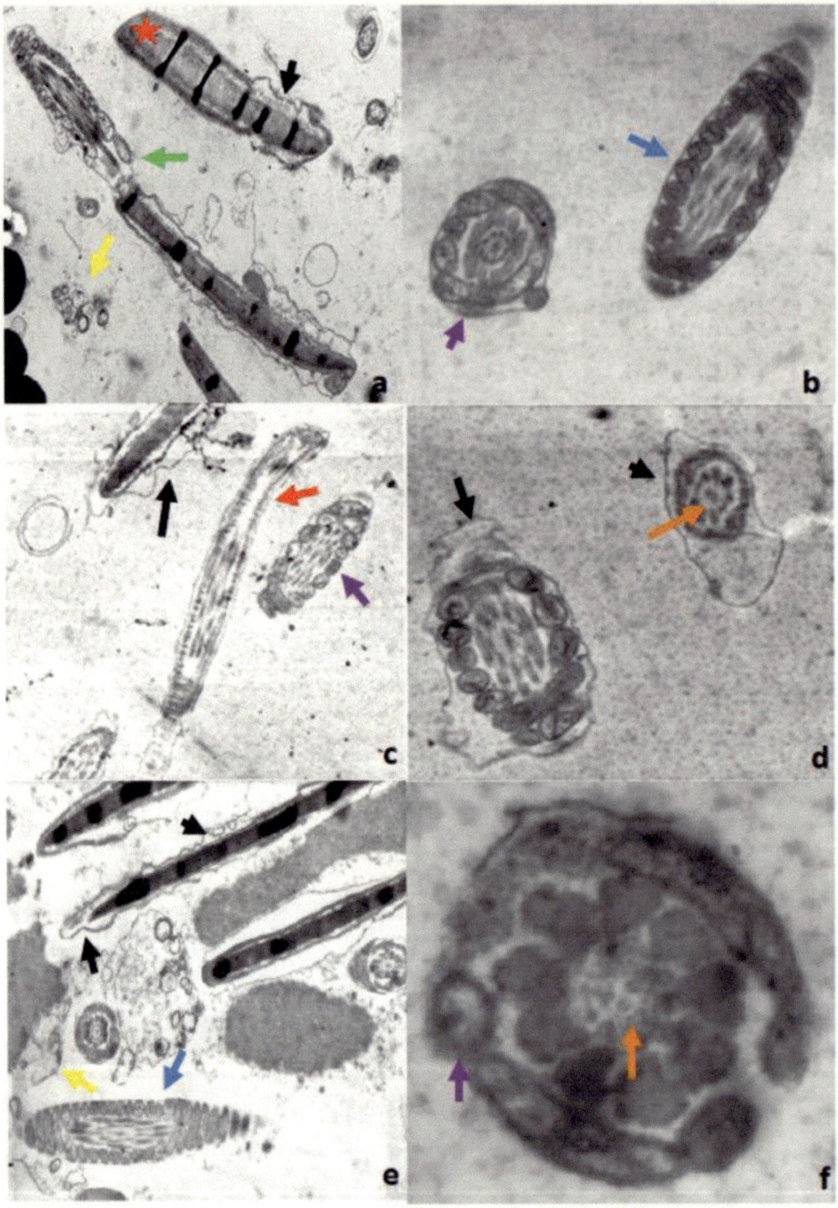

Figure 20.
STEM micrographs of frozen-thawed spermatozoa, (a) nucleus (red star), membrane damages (black arrow), neck damage (green arrow), membrane fragments (yellow arrow). (b) Mithochondrial damage (purple arrow), undamaged middle region (blue arrow). (c) Membrane damages (black arrow), mitochondrial damage (purple arrow), damaged middle region (red arrow). (d) Membrane damages (black arrow), axoneme damage (orange arrow). (e) Membrane damages (black arrow), membrane fragments (yellow arrow), undamaged middle region (blue arrow). (f) Undamaged middle region (blue arrow), axoneme damage (orange arrow) [88, 89].

intracellular ice formation and recrystallization has been demonstrated in yeast, plant cells [91], hamster tissue culture cells [92], and turkey sperm [93]. During defrosting, recrystallization generally starts at −40°C, and it becomes intense at the temperature range of −25 to −20°C.

The recrystallization mechanism consists of three types, isomass, active, and migratory.

1. *Isomass*: Isomass is a form of recrystallization that changes the shape of ice crystal. It can also be defined as the tendency of uneven rough surfaced or protruding ice particles combined to form larger particles having a more oval and smooth surface (**Figure 21**).

Figure 21.
Schematic drawing of isomass recrystallization.

2. *Active*: In the active recrystallization model, two or more smaller ice crystals combine to form a larger ice crystal (**Figure 22**).

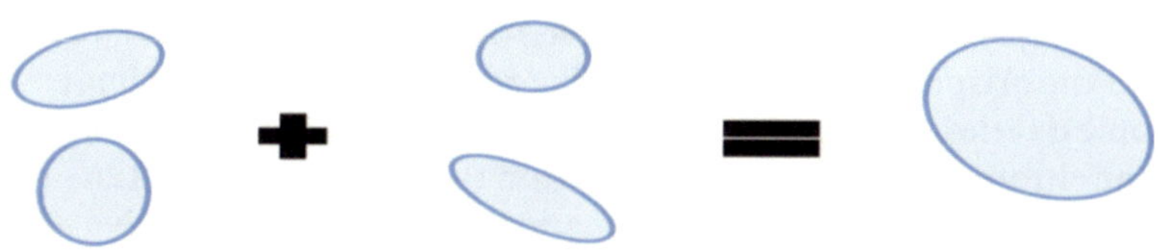

Figure 22.
Schematic drawing of active recrystallization.

3. *Migratory*: In the migratory recrystallization model, small crystals around a large crystal melt into a large crystal structure (**Figure 23**).

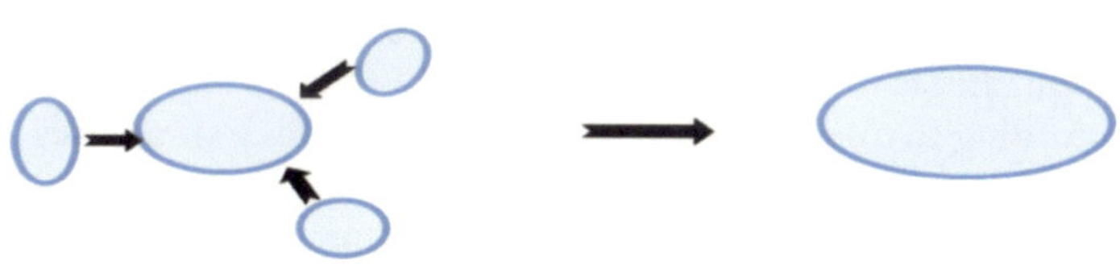

Figure 23.
Schematic drawing of migratory recrystallization [91].

Results of the previous studies revealed that the effect of ice formation or recrystallization depends on the crystal size [93] and on the total amount of ice in the cell [94]. In this context, rapid thawing protocol is more beneficial in terms of motility, viability, and membrane-organellar integrity, since small ice crystals do not have enough time to recrystallize when a fast warming protocol is applied. However, Mackenzie has shown that yeast cells cooled in optimal rate do not survive if they are quickly thawed, but sustain vitality when they are slowly thawed [92].

Another damage during thawing is due to the osmotic stress. This is especially true in frozen thawed rabbit [95], mouse [96], and bovine embryos and also freeze-thawed lymphocytes [97] and hematopoietic cells [16, 98]. In addition, remaining cryoprotective agents inside the cell are incorporated into the cytosolic water, lead to cell swollen, and cause various damage in the cell membrane, possibly cell lysis [99].

In the fertilization process, when the sperm reaches the fertilization zone, it enters between oviductal epithelial cells, infiltrates into the oocyte zona pellucida, undergoes acrosome reaction, opens a pore in the zone pellucida, and finally reaches oolemma. Disturbances in sperm membrane function due to cryopreservation directly affect this entire fertilization mechanism. In this case, fertilization cannot occur under vivo conditions. A spermatozoon with poor motility will have difficulty reaching the fertilization zone, hence has a poor ability to fertilize ovum.

It has been found that after the freeze-thaw procedure, human and bull spermatozoa contain more calcium than unprocessed samples, due to impairment of the selective-membrane permeability of the spermatozoa. Also, the affected spermatozoa cannot exert normal capacitation and fertilization due to high calcium concentration of the cytosolic environment, which also leads to premature acrosomal defects and sperm hyperactivation, causes losses in both capacitation and acrosome reaction [100].

7.4 The DNA damage

Spermatozoa with DNA damage have relatively limited fertilization ability [101]. These cells have fertilization defects and potentially transfer abnormal genetic materials to the offsprings. These genetic disorders might result in abnormality in embryonic development and aneuploidies (a numerical chromosomal disorder in which the chromosome number differs from the normal, diploid (2n) [102]. Because the spermatozoa with serious genetic damage have low in-vivo and in-vitro fertilization scores, a limited number of embryo could be harvested and embryos will have many of the developmental problems. In the light of these findings, it was concluded that sperm chromatin/DNA integrity is an important prerequisite for healthy fertilization and embryonic development [103]. In addition to abnormal DNA/nuclear protein interactions arising from intracellular ice crystals [104], scientists attributed the DNA damage during freeze-thaw processes to two different mechanisms:

The first mechanism proposed by Zribi et al. [105] is due to DNA damage in cryopreservation, and is thought to be caused by the activation of caspases and apoptosis. According to this hypothesis, apoptotic body-like structures are formed in the sperm cells during the freezing process. Apoptotic changes show themselves as nuclear fragmentation, chromatin condensation, mitochondrial expansion, and unusual changes in the cell membrane [106]. These changes produce the enzymes from cysteine protease group, called caspases, which play an important role during apoptosis. These enzymes are found in the cytoplasm as pro-caspases in zymogen granules and their activation requires a two-step cleavage in the specific aspartic acid residues to form active holoenes [9]. The caspases activate sequentially (waterfall-like reaction sequence), leading to a proteolytic helmet. Initiator caspases transduce death signals initiated by apoptotic stimulation to effector caspases. The effector caspases are related proteins, for example, actin or fosdrin from cytoskeletal proteins, lamin-A, the nuclear membrane protein, and poly (ADP-ribose) polymerase (PARP) involved in DNA repair break down apoptotic cell morphology [8]. DNA fragmentation is the latest sign of apoptosis, which occurs due to caspase-3 activation. The caspase-3 enzyme inhibits DNA repair by inactivating DNA repairing enzymes namely poly ADP-ribose polymerase (PARP) **Table 2**.

Current studies have shown that there is a positive correlation between the presence of PARP protein and sperm maturation. PARP homologs PARP-1 (75 kDa), PARP-9 (63 kDa), and PARP-2 (60 kDa) were found in injected sperm [107]. In another study [108], oxidative stress and PARP inhibition occurred in early

Subgroup	Function	Member
I	Activator (initiator caspases)	Caspase 2
		Caspase 8
		Caspase 9
		Caspase 10
II	Executioner (Effector caspases)	Caspase 3
		Caspase 6
		Caspase 7
III	Inflammatory mediators	Caspase 1
		Caspase 4
		Caspase 5
		Caspase 11
		Caspase 12
		Caspase 13
		Caspase 14

Table 2.
Caspase family subgroups and their members [8].

apoptotic spermatozoa, with water leakage, cell shrinking, and picnosis (volume reduction of the cell nucleus and chromatin condensation). The apoptosis-inducing factor released by the mitochondria (flavin), an enzyme that contains adenine dinucleotide helps oxidization during oxidative phosphorylation, causes DNA fragmentation in flavoprotein [109, 110]. During freezing and thawing, an increase in apoptosis markers was observed. Changes in the apoptotic membrane permeability, membrane potential, caspase activation, and phosphatidylserine externalization due to cryopreservation are also observed [111].

The second mechanism proposed by Thomson et al. [112] is that DNA damage is caused by caspase activation and oxidative stresses in the apoptotic cells exposed to extreme hostility. It is highly possible that freeze-thaw process results in the formation of free oxygen radicals that damage the nuclear DNA [113, 114]. With the weakened DNA repair mechanism [115], the frozen-thawed sperms are vulnerable to oxidative attacks [116]. An increase in the mitochondrial membrane potential due to changes in its membrane fluidity during sperm cryopreservation was also observed, which leads to the formation of free oxygen radicals. Accordingly, DNA damage to the spermatozoon single and double strand breaks by the released free oxygen radicals are very common. The formation of free oxygen radicals reaches peak level in human sperm and seminal leukocytes at 4°C [117]. Frozen semen samples containing high percentage of sperm are more prone to DNA damage. Although the cryopreservation adversely affects antioxidant activity, interestingly, it makes spermatozoa less vulnerable to free oxygen radical damage [118].

The DNA damage in the cryopreserved spermatozoa mainly occurs at early stages during/after thawing. In a previous experiment [119], the highest DNA damage score was observed within the first 4 h after thawing. Therefore, freeze-thawed semen samples should be used immediately in the clinical settings [119]. However, Isachenko et al. [120] suggested that freezing or cryoprotectant did not make any difference in DNA integrity in their study **Table 3**.

References	Cryopreservation method	Results
Spano et al. [122], Donelly et al. [123], Gandini et al. [124], Zribi et al. [105]	Equilibration at 37°C, freezing in liquid nitrogen at −80°C, storage in liquid nitrogen at −196°C	The freezing/thawing procedure causes DNA damage in the sperm.
De Paula et al. [125]	Freezing up to −20°C, freezing in liquid nitrogen, storage in liquid nitrogen at −196°C	
Petyim and Choavaratana [126]	Freezing with liquid nitrogen under computer control	
Ngamwuttiwong and Kunathikom [127]	Freezing in liquid nitrogen vapor	
Thomson et al. [112]	Programmable freezer	
Steele et al. [128]	Freezing in liquid nitrogen vapor	
Duru et al. [129]	Equilibration at 37°C, freezing in liquid nitrogen at −80°C, storage in liquid nitrogen at −196°C	
Isachenko et al. [130]	Programmable slow cooler + vitrification	
Paasch et al. [131]	Cooling to −20°C, freezing in liquid nitrogen at −100°C, storage in liquid nitrogen at −196°C	

Table 3.
Different studies showing the effects of freezing/thawing protocols on DNA damage [121].

7.5 The mitochondrial damage

Mitochondria and other membrane-bound organelles are found in the majority of eukaryotic cells and the mitochondria have a vital role in ATP synthesis in cells via oxidative phosphorylation [132]. Mitochondrion is found in nine fibrous columns located in the body of the spermatozoon and serves as an energizing engine for the semen [133]. Plasma membranes and mitochondrial membranes show similar sensitivity to cryopreservation. High levels of cholesterol and polyunsaturated fatty acids cause the membrane to become more fluidic under cold conditions.

Mitochondrion is unique double membrane-bound organelle in the cell. There is an intermembraneous space between the inner and outer mitochondrial membranes. Matrix facing surface of the inner membrane have oxidative-phosphorylation enzyme system and the cytoplasmic part consisting of structural and functional subunits in which many enzymes are organized. These proteins are crucial for the metabolic processes required by the cell and for the maintenance of the cellular structure. Morphologically, the mitochondria show different characteristics in the sperm of different animal species. While in humans, rhesus monkey, dogs and other mammals, the mitochondria are in oval form, whereas they are cylindrical in fish [134]. The size and number of mitochondria in a given cell may vary according to the bioenergetic needs of species. Mitochondrion has its own DNA (mtRNA) involved in the synthesis of messenger RNA, ribosomal RNA, and transfer RNA, which are highly needed for the internal metabolism of the cell [135]. In the mammalian spermatozoa, mitochondria are located at the middle part of the cell and will form multiple disulfide bonds to form the mitochondrial capsule [136, 137].

Sperm mitochondria, unlike those of the somatic cells, include cytochrome c, cytochrome c oxidase subunits, protein isoforms, and isoenzymes. Mitochondria play a very fundamental role in ATP production by oxidative phosphorylation. In the production of mitochondrial ATP, glycolysis uses carbon fuels such as pyruvate, glutamine, and amino acids [135]. At the same time, mitochondria are involved in other processes such as production of free radicals and apoptosis. The release of free oxygen radicals such as superoxide anion, hydrogen peroxide, and nitric oxide is important. If the free oxygen radical release exceeds the capacity of the antioxidant defense system of a cell, peroxidation of the biological membranes, especially sperm plasma membrane and undesirable conditions can induce oxidative stress, DNA-related aging, and apoptosis [138]. Most importantly, the mitochondria create the activation energy required for the movement of the sperm [139].

ATP formed by oxidative phosphorylation in mitochondria allows motility through microtubules. For this reason, any disorder in the mitochondrion metabolism leads to decreased motility. Oxidative phosphorylation in the mitochondrion requires two basic components in the inner membrane of the mitochondrion, the ATP synthase, and the respiratory chain [140].

During cryopreservation, two types of damage occur in mitochondria. The first is the direct damage in the mitochondrial DNA, deteriorations in the inner and outer membranes of the mitochondrion. The second is the indirect damage caused by losing ability to carry out genetic coding for mitochondrial activity [141]. According to Irvine et al. [142], motility and velocity decrease in sperm with high DNA damage. The nuclear DNA segments damaged during freeze-thawing process may be responsible for the loss of mitochondrial functions.

A change in the mitochondrial membrane fluidity also causes changes in the membrane potential and in exchange and release of free oxygen radicals through the mitochondrial membrane [107]. Peroxidative damage caused by an increase in the concentration of free oxygen radicals is associated with sperm plasma membrane damage and disruption of the aczonemal structure [143].

7.6 Free oxygen radicals and lipid peroxidation damage

Cryobiological events that occur during freezing not only cause physical damage but also disrupt the chemical structures. The cooling of the cell results in the production of free radicals, and it also suppresses the natural defensive mechanism that the cell develops against the formation of free radicals.

Depending on their low activation energy, the free oxygen radicals are more difficult to be removed at low temperatures than the normal enzymatic cleansing. Increasing amounts of free oxygen radicals affect various biomolecules including membrane lipids, proteins, and nucleic acids through nonenzymatic pathways and cause the formation of new free oxygen radicals in the environment. Transition metals such as copper and iron, which are highly effective in intracellular metabolism, such as the cytochromes of the mitochondrial electron transport chain, are important catalysts for the radical chain reactions. In particular, when catalysts are present in the species capable of redox cycling, they exchange electrons mutually between the oxidation sites.

The disruption of the homeostatic balance of the metal inlet and outlet of the cell during cooling leads to the formation of free oxygen radicals, damages the cell wall by thickening. Under normal conditions, while the formation and decomposition of free oxygen radicals are in equilibrium, the disruption of this equilibrium causes continuous formation and association of free oxygen radicals. Restricted free oxygen radicals cause damage to the cellular membrane by impairing lipid peroxidation. Under normal conditions, the cells have ways to repair such damage, but irreversible damage during storage and thawing can lead to cell death.

During irreversible cell damage or cell death, calcium is an important agent in the cell necrosis. Due to the cold shock, unbalanced free radical formation causes an increase in intracellular calcium. All of these events have increased the importance of antioxidants and calcium channel blockers (calcium enters and exits through the calcium channels in the membrane from the extracellular medium, and this entry-exit mechanism is controlled by the so-called calcium channel blockers) during freezing and thawing. These blockers consist of three prototype agents, dilithazem, verapamil, and nifedipine, and many sub-agents such as amlodipine, benidipine, cilnidipine, felodipine, isradipine, nilvadipine, etc.

Free oxygen radicals are separated into three groups;

1. *Oxygen free radicals*: superoxide anions (O_2), hydroxyl radical (OH), and hyperoxyl radical (HOO).

2. *Non-radical species*: hypochloric acid (HOCl) and hydrogen peroxide (H_2O_2).

3. *Reactive nitrogen species and free nitrogen radicals*: nitroxylon, nitrogen oxide, peroxynitrite, etc. [144].

Chemically, radicals are molecules having one or more unpaired electrons. This chemical state causes an excessive amount of reactivity with an electrically irregular structure in the predominant molecules. As in the other cells, energy is also produced aerobically in the spermatozoa. Energy is produced by oxidative phosphorylation in mitochondria and by oxidation of nicotinamide adenine dinucleotide (NADH) and is stored in the form of adenosine triphosphate (ATP). In this whole electron transfer chain, oxygen is reduced to free radicals in high electroactivity by taking four electrons. As a result of this reduction reaction, water becomes clear (**Figure 24**).

7.6.1 Lipid peroxidation

Reactive oxygen groups, which occur during the electron transport chain in the mitochondria, act on membrane lipids, rich in unsaturated fatty acids, causing deterioration of their structure. This is called lipid peroxidation.

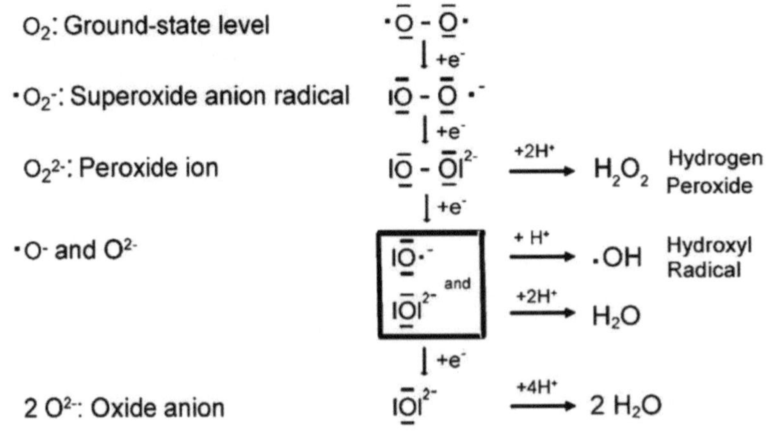

Figure 24.
Formation of reactive oxygen species [145].

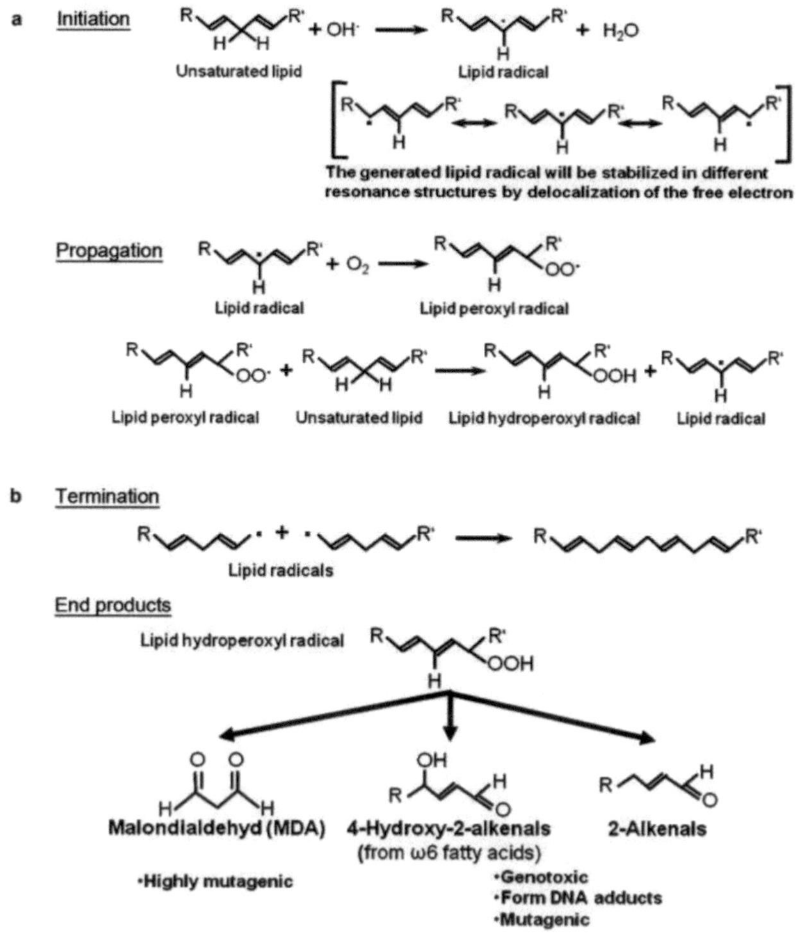

Figure 25.
Formation stages, propagation, and end products of lipid peroxidation [145].

Lipid peroxidation is exerted in three phases, initiation, propagation, and termination:

Initiation phase: In the initial phase, the highly reactive OH^- or hydroxyperoxyl radical ($-HO_2$) removes a proton (H^+) from the medium by affecting the polyunsaturated fatty acids in the lipid layer and converts their existing structure into a lipid radical and releases water. Low reactivity oxygen groups such as H_2O_2 cannot initiate this reaction [146]. In the present case, lipid radicals can be in various forms with electrons in delocalized state. At this stage, lipid radicals can react with molecular oxygen and are converted to lipid peroxides.

Propagation phase: During the propagation phase, the lipid radical formed in the previous phase merges with the oxygen in the molecular state and forms the lipid peroxide radical. The lipid peroxide radical reacts with a neighboring fatty acid to form a new lipid peroxidase and becomes itself a lipid hydroperoxyl. The ongoing reactions in this way are also referred to as the "radical chain reaction." The lipid peroxidation event affects 60% of the membrane lipids [147].

Termination phase: In the last phase, two or more lipid radicals bind to each other to form hydroperoxyl radicals such as malandialdehyde (MDA), 4-hydroxy-2-alkene, and 2-alkenal, which have dangerous properties in the last stage (**Figure 25**).

8. Antioxidants as cellular defense mechanism

In the previous sections, we mentioned about the free oxygen radicals formed in the cell and the antioxidants those produced by the cells to protect themselves from diverse effects of free radicals.

Antioxidants exert their effects by:

- Preventing free oxygen radicals from starting chain reactions,
- Stopping chain reactions that have begun,
- Preventing the formation of free oxygen radicals, and
- Performing peroxides by breaking down and reducing local oxygen concentration [148].

Under normal conditions, spermatozoa reduce antioxidants to counteract with free oxygen radicals that are produced metabolically. These molecules are mainly enzymes. The most known antioxidant enzymes are superoxide dismutase (converts the superoxide anion to H_2O_2), glutathione peroxidase (detoxifies the organic peroxides), and catalase (converts H_2O_2 to water) [149, 150]. However, an abnormally high amount of free oxygen radicals released by the freeze-thaw procedure, damage the cell by overcoming its antioxidant defense capacity. In such cases, different antioxidants are supplied into the diluent solutions [151].

Antioxidants are divided into primary and secondary antioxidants.

Primary antioxidants: The antioxidants which blocks the chain reaction by cutting in the half enter this group. Antioxidants of this group are mainly: tocopherols (vitamin E), butyl-hydroxyl-anisole (BHA), butyl-hydroxy-toluene (BHT), and ethoxiquine.

Secondary antioxidants: These kinds of *antioxidants* prevent oxidation of lipids by delaying their oxidation. They act by binding to metal ions, which catalyze the oxidation of lipids, by binding oxygen itself, and by absorbing UV rays. Some antioxidants such as citric acid, amino acids, ethylene diamine tetra acetic acid (EDTA) and certain phosphoric acid derivatives, ascorbic acid (Vitamin C), ascorbyl palmitate, sulfites, erythorbic acid and sodium erythorbate, and glucose oxidase are examples of secondary antioxidants [152]. Bucak et al. obtained better spermatological parameters in the Ankara goat [153–155] and bull [156] spermatozoa frozen in the antioxidants added to the dilution solutions after thawing.

9. Discussion

The membrane structure in the semen cells is complex. Several types of phospholipids and proteins, cholesterol and oligosaccharides are involved in cell membrane structure in a manner appropriate for the cell structure and are effective in providing intracellular and extracellular balance.

The osmotic balance between the intracellular and extracellular environment, DNA integrity, the mitochondrial energy system, and many other systematic ovum fertilization by spermatozoa and the transfer of genetic material are all due to the complete and orderly functioning of this biological system as a whole. Although freezing and storing the semen samples at −196°C has been a groundbreak in artificial insemination, despite all the precautions taken, a slight decrease in the motility and vitality caused by cold shock occurred during freezing procedure.

Rapid cooling of the cells disrupts the membrane structure, and cells are subjected to physical pressure from both inside and outside due to the formation of intracellular and extracellular ice. Although the slow cooling of the environment removes the physical impression of the ice, the cells are trapped in the solutes with a high ion concentration due to excessive water loss. For this reason, they are exposed to chemical deterioration.

To avoid these two situations, scientists have made efforts to determine the optimum cooling rate, and finally, optimal cooling rates have been determined for the semen of different mammalian species. In addition, various cryoprotectants have been used to prevent ice formation and deterioration of membrane integrity, but these substances with various side effects have also damaged the cells when supplemented.

DNA, which has a highly complex structure, is damaged even when the cooling rate and cryoprotectant are applied at optimal levels. The free oxygen radicals, which are formed as a result of slow metabolism, terrorize the cellular milieu, deteriorate the metabolism, and disrupt almost all functions of the cell. Against these toxic agents, cells produce protective mechanisms called antioxidants, but they are ineffective against high concentrations of free radicals, which are formed excessively in the cold environment. Artificial antioxidants added into the solutions do not provide the desired success so far. Sperms can also be damaged due to recrystallization during thawing.

10. Conclusion

In this chapter, we provide detailed information about cryobiology and cryopreservation of sperm. Necrosis and apoptosis of the cell during freezing are described. The structure of the biological membranes is detailed, the damages of the membranes during the cryopreservation process are mentioned and some cryoprotectants used against membrane damages are given. The structure of water, the mechanism of freezing and ice formation, and the effect of intracellular/extracellular water presence on cryopreservation were explained. These data were enriched with electron microscope images. Apart from physical damage, DNA damage, mitochondrial damage, ROS formation, lipid peroxidation damage, and antioxidative defense mechanism are mentioned. A comprehensive section on cryobiology and freezing of spermatozoa has been tried to be prepared, and it is aimed to give detailed information about cryopreservation.

Author details

Ali Erdem Öztürk[1], Mustafa Numan Bucak[1], Mustafa Bodu[1], Nuri Başpınar[2], İlhami Çelik[3], Zhiquan Shu[4,5], Nazan Keskin[6] and Dayong Gao[4*]

1 Department of Reproduction and Artificial Insemination, Selcuk University, Konya, Turkey

2 Department of Biochemistry, Selcuk University, Konya, Turkey

3 Department of Histology and Embriyology, Selcuk University, Konya, Turkey

4 Department of Mechanical Engineering, University of Washington, Seattle, WA, USA

5 School of Mechanical and Materials Engineering, Washington State University, Everett, WA, USA

6 Department of Histology and Embryology, Pamukkale University, Denizli, Turkey

*Address all correspondence to: dayong@uw.edu

References

[1] Spallanzani L, Di Pietro P. Edizione nazionale delle opere di Lazzaro Spallanzani: Charles Bonnet. Mucchi; 1776

[2] Pesch S, Bergmann M. Structure of mammalian spermatozoa in respect to viability, fertility and cryopreservation. Micron. 2006;**37**:597-612

[3] Polge C, Smith AU, Parkes A. Revival of spermatozoa after vitrification and dehydration at low temperatures. Nature. 1949;**164**:666

[4] Bucak MN, Tekin N. Kryoprotektanlar ve gamet hücrelerinin dondurulmasında kryoprotektif etki. Ankara Üniversitesi Veteriner Fakültesi Dergisi. 2007;**54**:67-72

[5] Whittingham D, Leibo S, Mazur P. Survival of mouse embryos frozen to −196 and −269°C. Science. 1972;**178**:411-414

[6] Gao D, Critser J. Mechanisms of cryoinjury in living cells. ILAR Journal. 2000;**41**:187-196

[7] Benson JD, Woods EJ, Walters EM, Critser JK. The cryobiology of spermatozoa. Theriogenology. 2012;**78**:1682-1699

[8] Ozaydin T, Celik I. Histological, histochemical and immunohistochemical investigations on the developing small intestines of broilers embryos. Journal of Animal and Veterinary Advances. 2012;**11**:2936-2944

[9] Samali A, Orrenius S. Heat shock proteins: Regulators of stress response and apoptosis. Cell Stress and Chaperones. 1998;**3**:228

[10] Wyllie A, Duvall E. Cell injury and death. In: McGee JO'D, Isaacson PG, Nicholas A, Heather WM, Mary DPE, editors. Oxford Textbook of Pathology. Principles of Pathology. Vol. 1. Oxford, UK: Oxford University Press; Slack. 1992. pp. 141-193

[11] Tinaztepe K, Ozen S, Güçer S, Ozdamar S. Apoptosis in renal disease: A brief review of the literature and report of preliminary findings in childhood lupus nephritis. The Turkish Journal of Pediatrics. 2001;**43**:133-138

[12] Abastado J-P. Apoptosis: Function and regulation of cell death. Research in Immunology. 1996;**147**:443-456

[13] Gabai VL, Meriin AB, Mosser DD, Caron A, Rits S, Shifrin VI, et al. Hsp70 prevents activation of stress kinases a novel pathway of cellular thermotolerance. Journal of Biological Chemistry. 1997;**272**:18033-18037

[14] Nagle WA, Soloff BL, Moss A, Henle KJ. Cultured Chinese hamster cells undergo apoptosis after exposure to cold but nonfreezing temperatures. Cryobiology. 1990;**27**:439-451

[15] Elkind MM, Whitmore GF. Radiobiology of Cultured Mammalian Cells. New York, London: Gordon and Breach; 1967

[16] Mazur P. Freezing of living cells: Mechanisms and implications. American Journal of Physiology-Cell Physiology. 1984;**247**:C125-C142

[17] Singer S, Nicolson GL. The fluid mosaic model of the structure of cell membranes. Science. 1972;**175**:720-735

[18] Parks JE, Graham JK. Effects of cryopreservation procedures on sperm membranes. Theriogenology. 1992;**38**:209-222

[19] Holt W. Basic aspects of frozen storage of semen. Animal Reproduction Science. 2000;**62**:3-22

[20] Cabrita E, Sarasquete C, Martínez-Páramo S, Robles V, Beirão J, Pérez-Cerezales S, et al. Cryopreservation of fish sperm: Applications and perspectives. Journal of Applied Ichthyology. 2010;**26**:623-635

[21] Simpson A, Swan M, White I. Susceptibility of epididymal boar sperm to cold shock and protective action of phosphatidylcholine. Molecular Reproduction and Development. 1987;**17**:355-373

[22] Sariozkan S, Tuncer P, Bucak M, Buyukleblebici S, Kinet H, Bilgen A. The effects of different egg yolk concentrations used with soy bean lecithin-based extender on semen quality to freeze bull semen. Eurasian Journal of Veterinary Sciences. 2010;**26**:45-49

[23] Brill JA, Yildirim S, Fabian L. Phosphoinositide signaling in sperm development. Seminars in Cell and Developmental Biology. 2016;**59**:2-9

[24] Luconi M, Marra F, Gandini L, Filimberti E, Lenzi A, Forti G, et al. Phosphatidylinositol 3-kinase inhibition enhances human sperm motility. Human Reproduction. 2001;**16**:1931-1937

[25] de Vantéry Arrighi C, Lucas H, Chardonnens D, De Agostini A. Removal of spermatozoa with externalized phosphatidylserine from sperm preparation in human assisted medical procreation: Effects on viability, motility and mitochondrial membrane potential. Reproductive Biology and Endocrinology. 2009;7:1

[26] Wilhelm K, Graham J, Squires E. Effects of phosphatidylserine and cholesterol liposomes on the viability, motility, and acrosomal integrity of stallion spermatozoa prior to and after cryopreservation. Cryobiology. 1996;**33**:320-329

[27] Yeagle PL. The Membranes of Cells. London: Academic Press; 2016

[28] Medeiros CMO, Forell F, Oliveira ATD, Rodrigues JL. Current status of sperm cryopreservation: Why isn't it better? Theriogenology. 2002;**57**:327-344

[29] Van Dijck P, De Kruijff B, Van Deenen L, De Gier J, Demel R. The preference of cholesterol for phosphatidylcholine in mixed phosphatidylcholine-phosphatidylethanolamine bilayers. Biochimica et Biophysica Acta (BBA)—Biomembranes. 1976;**455**:576-587

[30] Yeagle PL. Cholesterol and the cell membrane. Biochimica et Biophysica Acta (BBA)—Reviews on Biomembranes. 1985;**822**:267-287

[31] Flesch FM, Brouwers JF, Nievelstein PF, Verkleij AJ, van Golde LM, Colenbrander B, et al. Bicarbonate stimulated phospholipid scrambling induces cholesterol redistribution and enables cholesterol depletion in the sperm plasma membrane. Journal of Cell Science. 2001;**114**:3543-3555

[32] Visconti PE, Ning X, Fornés MW, Alvarez JG, Stein P, Connors SA, et al. Cholesterol efflux-mediated signal transduction in mammalian sperm: Cholesterol release signals an increase in protein tyrosine phosphorylation during mouse sperm capacitation. Developmental Biology. 1999;**214**:429-443

[33] Chapman D. Phase transitions and fluidity characteristics of lipids and cell membranes. Quarterly Reviews of Biophysics. 1975;**8**:185-235

[34] Zhmakin AI. Fundamentals of Cryobiology: Physical Phenomena and Mathematical Models. Berlin: Springer; 2008

[35] Yeung C-H. Aquaporins in spermatozoa and testicular germ cells: Identification and potential role. Asian Journal of Andrology. 2010;**12**:490-499

[36] Seddon J, Templer R. Polymorphism of lipid-water systems. In: Handbook of Biological Physics. North Holland; Vol. 1. 1995. pp. 97-160

[37] Killian JA, de Kruijff B. Nonbilayer lipids affect peripheral and integral membrane proteins via changes in the lateral pressure profile. Biochimica et Biophysica Acta (BBA)—Biomembranes. 2004;**1666**:275-288

[38] Cullis PR, De Kruijff B. Lipid polymorphism and the functional roles of lipids in biological membranes. Biochimica et Biophysica Acta (BBA)—Reviews on Biomembranes. 1979;**559**:399-420

[39] Quinn P. A lipid-phase separation model of low-temperature damage to biological membranes. Cryobiology. 1985;**22**:128-146

[40] Hammerstedt R, Graham JK, Nolan JP. Cryopreservation of mammalian sperm: What we ask them to survive. Journal of Andrology. 1990;**11**:73-88

[41] Caldarelli G, D e Los Rios P. Cold and warm denaturation of proteins. Journal of Biological Physics. 2001;**27**:229-241

[42] Farrant J, Molyneux P, Hasted J, Meares P, Echlin P. Water transport and cell survival in cryobiological procedures [and discussion]. Philosophical Transactions of the Royal Society of London B: Biological Sciences. 1977;**278**:191-205

[43] Royere D, Barthelemy C, Hamamah S, Lansac J. Cryopreservation of spermatozoa: A 1996 review. Human Reproduction Update. 1996;**2**:553-559

[44] Crowe J, Crowe L. Effects of dehydration on membranes and membrane stabilization at low water activities. Biological Membranes. 1984;**5**:57-103

[45] Woelders H. Fundamentals and recent development in cryopreservation of bull and boar semen. Veterinary Quarterly. 1997;**19**:135-138

[46] Alvarenga MA, Papa FO, Landim-Alvarenga F, Medeiros A. Amides as cryoprotectants for freezing stallion semen: A review. Animal Reproduction Science. 2005;**89**:105-113

[47] Sağırkaya H, Bağış H. Memeli embriyolarının kriyoprezervasyonu. Uludag University Journal of the Faculty of Veterinary Medicine. 2003;**22**:127-135

[48] Purdy PH. A review on goat sperm cryopreservation. Small Ruminant Research. 2006;**63**:215-225

[49] Amann R, Pickett B. Principles of cryopreservation and a review of cryopreservation of stallion spermatozoa. Journal of Equine Veterinary Science. 1987;7:145-173

[50] Anchordoguy TJ, Rudolph AS, Carpenter JF, Crowe JH. Modes of interaction of cryoprotectants with membrane phospholipids during freezing. Cryobiology. 1987;**24**:324-331

[51] Armitage W. Osmotic stress as a factor in the detrimental effect of glycerol on human platelets. Cryobiology. 1986;**23**:116-125

[52] Niedermeyer W, Parish G, Moor H. Reactions of yeast cells to glycerol treatment alterations to membrane structure and glycerol uptake. Protoplasma. 1977;**92**:177-193

[53] Kachar B, Reese TS. Rapid formation of gap-junction-like structures induced by glycerol. The Anatomical Record. 1985;**213**:7-15

[54] Keates RA. Effects of glycerol on microtubule polymerization kinetics. Biochemical and Biophysical Research Communications. 1980;**97**:1163-1169

[55] Hammerstedt RH, Graham JK. Cryopreservation of poultry sperm: The enigma of glycerol. Cryobiology. 1992;**29**:26-38

[56] Gilmore J, McGann L, Liu J, Gao D, Peter A, Kleinhans F, et al. Effect of cryoprotectant solutes on water permeability of human spermatozoa. Biology of Reproduction. 1995;**53**:985-995

[57] Si W, Zheng P, Li Y, Dinnyes A, Ji W. Effect of glycerol and dimethyl sulfoxide on cryopreservation of rhesus monkey (*Macaca mulatta*) sperm. American Journal of Primatology. 2004;**62**:301-306

[58] Awad M. Effect of some permeating cryoprotectants on CASA motility results in cryopreserved bull spermatozoa. Animal Reproduction Science. 2011;**123**:157-162

[59] Pope D, Oliver W. Dimethyl sulfoxide (DMSO). Canadian Journal of Comparative Medicine and Veterinary Science. 1966;**30**:3

[60] Strong DM, Ahmed AA, Sell KW, Greiff D. In vitro effects of cryoprotective agents on the response of murine T and B lymphoid subpopulations to mitogenic agents. Cryobiology. 1972;**9**:450-456

[61] Trounson A, Mohr L. Human pregnancy following cryopreservation, thawing and transfer of an eight-cell embryo. Nature. 1983;**305**:707-709

[62] Ashwood-Smith M. Current concepts concerning radioprotective and cryoprotective properties of dimethyl sulfoxide in cellular systems. Annals of the New York Academy of Sciences. 1975;**243**:246-256

[63] Jones R. Collection, motility and storage of spermatozoa from the African elephant *Loxodonta africana*. Nature. 1973;**243**:38-39

[64] Rota A, Milani C, Cabianca G, Martini M. Comparison between glycerol and ethylene glycol for dog semen cryopreservation. Theriogenology. 2006;**65**:1848-1858

[65] Anchordoguy T, Crowe JH, Griffin FJ, Clark WH. Cryopreservation of sperm from the marine shrimp *Sicyonia ingentis*. Cryobiology. 1988;**25**:238-243

[66] Gwo J-C. Cryopreservation of black grouper (*Epinephelus malabaricus*) spermatozoa. Theriogenology. 1993;**39**:1331-1342

[67] Bucak MN, Tekin N. Protective effect of taurine, glutathione and trehalose on the liquid storage of ram semen. Small Ruminant Research. 2007;**73**:103-108

[68] Patist A, Zoerb H. Preservation mechanisms of trehalose in food and biosystems. Colloids and Surfaces B: Biointerfaces. 2005;**40**:107-113

[69] Ahmad E, Aksoy M. Trehalose as a cryoprotective agent for the sperm cells: A mini review. Animal Health, Production and Hygiene. 2012;**1**:123-129

[70] Hu JH, Li QW, Jiang ZL, Yang H, Zhang SS, Zhao HW. The cryoprotective effect of trehalose supplementation on boar spermatozoa quality. Reproduction in Domestic Animals. 2009;**44**:571-575

[71] Öztürk C, Güngör Ş, Ataman MB, Bucak MN, Başpinar N, Ili P, et al. Effects of arginine and trehalose on post-thawed bovine sperm quality. Acta Veterinaria Hungarica. 2017;**65**:429-439

[72] Uysal O, Bucak MN. The role of different trehalose concentrations and cooling rates in freezing of ram semen.

Ankara Universitesi Veteriner Fakultesi Dergisi. 2009;**56**:99-103

[73] Bucak MN, Ateşşahin A, Varışlı Ö, Yüce A, Tekin N, Akçay A. The influence of trehalose, taurine, cysteamine and hyaluronan on ram semen: Microscopic and oxidative stress parameters after freeze–thawing process. Theriogenology. 2007;**67**:1060-1067

[74] Benedict W, Gailar N, Plyler EK. Rotation-vibration spectra of deuterated water vapor. The Journal of Chemical Physics. 1956;**24**:1139-1165

[75] Finney JL. Water? What's so special about it? Philosophical Transactions of the Royal Society of London B: Biological Sciences. 2004;**359**:1145-1165

[76] Levy Y, Onuchic JN. Water mediation in protein folding and molecular recognition. Annual Review of Biophysics and Biomolecular Structure. 2006;**35**:389-415

[77] Oleinikova A, Brovchenko I, Smolin N, Krukau A, Geiger A, Winter R. Percolation transition of hydration water: From planar hydrophilic surfaces to proteins. Physical Review Letters. 2005;**95**: 247802

[78] Petukhov M, Rychkov G, Firsov L, Serrano L. H-bonding in protein hydration revisited. Protein Science. 2004;**13**:2120-2129

[79] Drobnis EZ, Crowe LM, Berger T, Anchordoguy TJ, Overstreet JW, Crowe JH. Cold shock damage is due to lipid phase transitions in cell membranes: A demonstration using sperm as a model. Journal of Experimental Zoology. 1993;**265**:432-437

[80] Mazur P. Principles of cryobiology. In: Life in the Frozen State. Boca Raton: CRC Press; 2004. pp. 3-65

[81] Mazur P, Koshimoto C. Is intracellular ice formation the cause of death of mouse sperm frozen at high cooling rates? 1. Biology of Reproduction. 2002;**66**:1485-1490

[82] Muldrew K. The salting-in hypothesis of post-hypertonic lysis. Cryobiology. 2008;**57**:251-256

[83] Lovelock J. The haemolysis of human red blood-cells by freezing and thawing. Biochimica et Biophysica Acta. 1953;**10**:414-426

[84] Henry MA, Noiles EE, Gao D, Mazur P, Critser JK. Cryopreservation of human spermatozoa. IV. The effects of cooling rate and warming rate on the maintenance of motility, plasma membrane integrity, and mitochondrial function. Fertility and Sterility. 1993;**60**:911-918

[85] Fiser P, Fairfull R. Combined effect of glycerol concentration and cooling velocity on motility and acrosomal integrity of boar spermatozoa frozen in 0.5 ml straws. Molecular Reproduction and Development. 1990;**25**:123-129

[86] Duncan AE, Watson P. Predictive water loss curves for ram spermatozoa during cryopreservation: Comparison with experimental observations. Cryobiology. 1992;**29**:95-105

[87] Morris G. Rapidly cooled human sperm: No evidence of intracellular ice formation. Human Reproduction. 2006;**21**:2075-2083

[88] Bucak NKM, Erdoğan C, İli AP, Başpınar N, Dursun Ş, Güngör Ş, et al. Effects of cryoprotectants and trehalose on electron microscopic evaluation of cryopreserved semen. Journal of Reproduction and Infertility. 2017;**18**:25-26

[89] Altuntaş CE. Ultrastructural investigation of the cryoprotective

effects of trehalose on frozen-thawed ram spermatozoa [Msc thesis]. Denizli, Turkey: Department of Histology and Embryology, Pamukkale University Graduate School of Health Sciences; 2018

[90] Lovelock JE. The denaturation of lipid-protein complexes as a cause of damage by freezing. Proceedings of the Royal Society of London, Series B: Biological Sciences. 1957;**147**:427-433

[91] Oude Vrielink AS, Aloi A, Olijve LL, Voets IK. Interaction of ice binding proteins with ice, water and ions. Biointerphases. 2016;**11**:018906

[92] MacKenzie A. Death of frozen yeast in the course of slow warming. In: Ciba Foundation Symposium-The Frozen Cell. London: Wiley Online Library; 1970. pp. 89-96

[93] Shimada K. Effects of cryoprotective additives on intracellular ice formation and survival in very rapidly cooled heLa cells. Contributions from the Institute of Low Temperature Science. 1977;**19**:49-69

[94] Farrant J, Walter C, Lee H, McGann L. Use of two-step cooling procedures to examine factors influencing cell survival following freezing and thawing. Cryobiology. 1977;**14**:273-286

[95] Bank H, Maurer R. Survival of frozen rabbit embryos. Experimental Cell Research. 1974;**89**:188-196

[96] Whittingham D. The viability of frozen-thawed mouse blastocysts. Journal of Reproduction and Fertility. 1974;**37**:159-162

[97] Thorpe P, Knight SC, Farrant J. Optimal conditions for the preservation of mouse lymph node cells in liquid nitrogen using cooling rate techniques. Cryobiology. 1976;**13**:126-133

[98] Crowder J, Dunn C, Jones J. Cryopreservation of erythropoietin-responsive cells in murine hematopoietic tissue. Cryobiology. 1980;**17**:18-24

[99] Gage AA, Baust J. Mechanisms of tissue injury in cryosurgery. Cryobiology. 1998;**37**:171-186

[100] Bailey JL, Bilodeau J, Cormier N. Semen cryopreservation in domestic animals: A damaging and capacitating phenomenon. Journal of Andrology. 2000;**21**:1-7

[101] Ardon F, Helms D, Sahin E, Bollwein H, Töpfer-Petersen E, Waberski D. Chromatin-unstable boar spermatozoa have little chance of reaching oocytes in vivo. Reproduction. 2008;**135**:461-470

[102] Evenson D. Loss of livestock breeding efficiency due to uncompensable sperm nuclear defects. Reproduction, Fertility and Development. 1999;**11**(1):16

[103] Fraser L, Strzeżek J, Kordan W. Effect of freezing on sperm nuclear DNA. Reproduction in Domestic Animals. 2011;**46**:14-17

[104] Royere D, Hamamah S, Nicolle J, Lansac J. Chromatin alterations induced by freeze-thawing influence the fertilizing ability of human sperm. International Journal of Andrology. 1991;**14**:328-332

[105] Zribi N, Chakroun NF, ElEuch H, Gargouri J, Bahloul A, Keskes LA. Effects of cryopreservation on human sperm deoxyribonucleic acid integrity. Fertility and Sterility. 2010;**93**:159-166

[106] Kerr JF, Wyllie AH, Currie AR. Apoptosis: A basic biological phenomenon with wide-ranging implications in tissue kinetics. British Journal of Cancer. 1972;**26**:239

[107] Said TM, Gaglani A, Agarwal A. Implication of apoptosis in sperm cryoinjury. Reproductive Biomedicine Online. 2010;**21**:456-462

[108] Mahfouz RZ, Sharma RK, Poenicke K, Jha R, Paasch U, Grunewald S, et al. Evaluation of poly (ADP-ribose) polymerase cleavage (cPARP) in ejaculated human sperm fractions after induction of apoptosis. Fertility and Sterility. 2009;**91**:2210-2220

[109] Candé C, Cohen I, Daugas E, Ravagnan L, Larochette N, Zamzami N, et al. Apoptosis-inducing factor (AIF): A novel caspase-independent death effector released from mitochondria. Biochimie. 2002;**84**:215-222

[110] Martin G, Cagnon N, Sabido O, Sion B, Grizard G, Durand P, et al. Kinetics of occurrence of some features of apoptosis during the cryopreservation process of bovine spermatozoa. Human Reproduction. 2007;**22**:380-388

[111] Martin G, Sabido O, Durand P, Levy R. Cryopreservation induces an apoptosis-like mechanism in bull sperm. Biology of Reproduction. 2004;**71**:28-37

[112] Thomson LK, Fleming SD, Aitken RJ, DeIuliis GN, Zieschang J-A, Clark AM. Cryopreservation-induced human sperm DNA damage is predominantly mediated by oxidative stress rather than apoptosis. Human Reproduction. 2009;**24**:2061-2070

[113] Baumber J, Ball BA, Linfor JJ, Meyers SA. Reactive oxygen species and cryopreservation promote DNA fragmentation in equine spermatozoa. Journal of Andrology. 2003;**24**:621-628

[114] Lewis S, Aitken R. DNA damage to spermatozoa has impacts on fertilization and pregnancy. Cell and Tissue Research. 2005;**322**:33-41

[115] Ahmadi A, Ng SC. Fertilizing ability of DNA-damaged spermatozoa. Journal of Experimental Zoology. 1999;**284**:696-704

[116] Rath D, Bathgate R, Rodriguez-Martinez H, Roca J, Strzezek J, Waberski D. Recent advances in boar semen cryopreservation. In: Control of Pig Reproduction VIII. Nottingham University Press, Nottingham; 2009. pp. 51-66

[117] Wang AW, Zhang H, Ikemoto I, Anderson DJ, Loughlin KR. Reactive oxygen species generation by seminal cells during cryopreservation. Urology. 1997;**49**:921-925

[118] Lasso JL, Noiles EE, Alvarez JG, Storey BT. Mechanism of superoxide dismutase loss from human sperm cells during cryopreservation. Journal of Andrology. 1994;**15**:255-255

[119] Gosálvez J, Cortés-Gutierez E, López-Fernández C, Fernández JL, Caballero P, Nuñez R. Sperm deoxyribonucleic acid fragmentation dynamics in fertile donors. Fertility and Sterility. 2009;**92**:170-173

[120] Isachenko V, Isachenko E, Katkov II, Montag M, Dessole S, Nawroth F, et al. Cryoprotectant-free cryopreservation of human spermatozoa by vitrification and freezing in vapor: Effect on motility, DNA integrity, and fertilization ability. Biology of Reproduction. 2004;**71**:1167-1173

[121] Di Santo M, Tarozzi N, Nadalini M, Borini A. Human sperm cryopreservation: Update on techniques, effect on DNA integrity, and implications for ART. Advances in Urology. 2012;**2012**:854837

[122] Spano M, Cordelli E, Leter G, Lombardo F, Lenzi A, Gandini L. Nuclear chromatin variations in human spermatozoa undergoing swim-up and cryopreservation evaluated by the flow

cytometric sperm chromatin structure assay. Molecular Human Reproduction. 1999;**5**:29-37

[123] Donnelly ET, McClure N, Lewis SE. Cryopreservation of human semen and prepared sperm: Effects on motility parameters and DNA integrity. Fertility and Sterility. 2001;**76**:892-900

[124] Gandini L, Lombardo F, Lenzi A, Spano M, Dondero F. Cryopreservation and sperm DNA integrity. Cell and Tissue Banking. 2006;**7**:91-98

[125] de Paula TS, Bertolla RP, Spaine DM, Cunha MA, Schor N, Cedenho AP. Effect of cryopreservation on sperm apoptotic deoxyribonucleic acid fragmentation in patients with oligozoospermia. Fertility and Sterility. 2006;**86**:597-600

[126] Petyim S, Choavaratana R. Cryodamage on sperm chromatin according to different freezing methods, assessed by AO test. Journal of The Medical Association of Thailand. 2006;**89**:306-313

[127] Ngamwuttiwong T, Kunathikom S. Evaluation of cryoinjury of sperm chromatin according to liquid nitrogen vapour method (I). Journal of the Medical Association of Thailand. 2007;**90**:224

[128] Steele EK, McClure N, Lewis SE. Comparison of the effects of two methods of cryopreservation on testicular sperm DNA. Fertility and Sterility. 2000;**74**:450-453

[129] Duru NK, Morshedi MS, Schuffner A, Oehninger S. Cryopreservation-thawing of fractionated human spermatozoa is associated with membrane phosphatidylserine externalization and not DNA fragmentation. Journal of Andrology. 2001;**22**:646-651

[130] Isachenko E, Isachenko V, Katkov II, Rahimi G, Schondorf T, Mallmann P, et al. DNA integrity and motility of human spermatozoa after standard slow freezing versus cryoprotectant-free vitrification. Human Reproduction. 2004;**19**:932-939

[131] Paasch U, Sharma RK, Gupta AK, Grunewald S, Mascha EJ, Thomas AJ Jr, et al. Cryopreservation and thawing is associated with varying extent of activation of apoptotic machinery in subsets of ejaculated human spermatozoa 1. Biology of Reproduction. 2004;**71**:1828-1837

[132] Devenish RJ, Prescott M, Rodgers AJ. The structure and function of mitochondrial F 1 F 0-ATP synthases. International Review of Cell and Molecular Biology. 2008;**267**:1-58

[133] O'connell M, McClure N, Lewis S. The effects of cryopreservation on sperm morphology, motility and mitochondrial function. Human Reproduction. 2002;**17**:704-709

[134] Lahnsteiner F, Patzner RA. Sperm morphology and ultrastructure in fish. In: Fish Spermatology. Oxford, UK: Alpha Science International Ltd; 2008. pp. 1-61

[135] Piomboni P, Focarelli R, Stendardi A, Ferramosca A, Zara V. The role of mitochondria in energy production for human sperm motility. International Journal of Andrology. 2012;**35**:109-124

[136] Calvin HI, Cooper GW, Wallace E. Evidence that selenium in rat sperm is associated with a cysteine-rich structural protein of the mitochondrial capsules. Gamete Research. 1981;**4**:139-149

[137] Ursini F, Heim S, Kiess M, Maiorino M, Roveri A, Wissing J, et al. Dual function of the selenoprotein PHGPx during sperm maturation. Science. 1999;**285**:1393-1396

[138] Figueroa E, Valdebenito I, Zepeda AB, Figueroa CA, Dumorné K, Castillo RL, et al. Effects of cryopreservation on mitochondria of fish spermatozoa. Reviews in Aquaculture. 2017;**9**:76-87

[139] Guthrie H, Welch G. Effects of reactive oxygen species on sperm function. Theriogenology. 2012;**78**:1700-1708

[140] Brown K. Fish mitochondrial genomics: Sequence, inheritance and functional variation. Journal of Fish Biology. 2008;**72**:355-374

[141] Kurland C, Andersson S. Origin and evolution of the mitochondrial proteome. Microbiology and Molecular Biology Reviews. 2000;**64**:786-820

[142] Irvine DS, Twigg JP, Gordon EL, Fulton N, Milne PA, Aitken R. DNA integrity in human spermatozoa: Relationships with semen quality. Journal of Andrology. 2000;**21**:33-44

[143] Saleh R, Agarwal A. Oxidative stress and male infertility: From research bench to clinical practice. Journal of Andrology. 2002;**23**:737

[144] Sabeti P, Pourmasumi S, Rahiminia T, Akyash F, Talebi AR. Etiologies of sperm oxidative stress. International Journal of Reproductive BioMedicine. 2016;**14**:231

[145] Henkel RR. Leukocytes and oxidative stress: Dilemma for sperm function and male fertility. Asian Journal of Andrology. 2011;**13**:43

[146] Blake D, Allen R, Lunec J. Free radicals in biological systems—A review orientated to inflammatory processes. British Medical Bulletin. 1987;**43**:371-385

[147] Kothari S, Thompson A, Agarwal A, du Plessis SS. Free radicals: Their beneficial and detrimental effects on sperm function. Indian Journal of Experimental Biology. 2010;**48**:425-435

[148] Cheeseman K, slater T. An introduction to free radical bio-chemistry. British Medical Bulletin. 1993;**49**:481-493

[149] Guemouri L, Artur Y, Herbeth B, Jeandel C, Cuny G, Siest G. Biological variability of superoxide dismutase, glutathione peroxidase, and catalase in blood. Clinical Chemistry. 1991;**37**:1932-1937

[150] Marti E, Marti J, Muiño-Blanco T, Cebrián-Pérez J. Effect of the cryopreservation process on the activity and immunolocalization of antioxidant enzymes in ram spermatozoa. Journal of Andrology. 2008;**29**:459-467

[151] Bucak MN, Tuncer PB, Sarıözkan S, Başpınar N, Taşpınar M, Çoyan K, et al. Effects of antioxidants on post-thawed bovine sperm and oxidative stress parameters: Antioxidants protect DNA integrity against cryodamage. Cryobiology. 2010;**61**:248-253

[152] Ömür AD. Koç Spermasının Dondurulmasında Antioksidanların Etkisi. Atatürk Üniversitesi Veteriner Bilimleri Dergisi. 2015;**10**:61-69

[153] Bucak MN, Sarıözkan S, Tuncer PB, Ulutaş PA, Akçadağ Hİ. Effect of antioxidants on microscopic semen parameters, lipid peroxidation and antioxidant activities in angora goat semen following cryopreservation. Small Ruminant Research. 2009;**81**:90-95

[154] Bucak MN, Tuncer PB, Sarıözkan S, Ulutaş PA, Çoyan K, Başpınar N, et al. Effects of hypotaurine, cysteamine and aminoacids solution on post-thaw microscopic and oxidative stress parameters of angora goat

semen. Research in Veterinary Science. 2009;**87**:468-472

[155] Bucak MN, Sarıözkan S, Tuncer PB, Sakin F, Ateşşahin A, Kulaksız R, et al. The effect of antioxidants on post-thawed angora goat (Capra hircus ancryrensis) sperm parameters, lipid peroxidation and antioxidant activities. Small Ruminant Research. 2010;**89**:24-30

[156] Bucak M, Ataman M, Başpınar N, Uysal O, Taşpınar M, Bilgili A, et al. Lycopene and resveratrol improve post-thaw bull sperm parameters: Sperm motility, mitochondrial activity and DNA integrity. Andrologia. 2015;**47**:545-552

Section 3

Cryobiology Aiding Organ Transplant

Chapter 6

Current Advancements in Pancreatic Islet Cryopreservation Techniques

Samuel Rodriguez, David Whaley, Michael Alexander, Mohammad Rezaa Mohammadi and Jonathan R.T. Lakey

Abstract

There have been significant advancements in the research of pancreatic islet transplantations over the past 50 years as a treatment for Type 1 Diabetes Mellitus (T1DM). This work has resulted in hundreds of clinical islet transplantation procedures internationally. One limitation of the procedure includes effective storage techniques during donor-recipient cross-matching following islet isolation from deceased donor. Cryopreservation, which is heavily used in embryology research, has been proposed as a prospective method for pancreatic islet banking to bridge the temporal intervals between donor-recipient matching. The cryopreservation methods currently involve the freezing of islets to subzero (−80/−196°C) temperatures for storage followed by a thawing and warming period, which can be increasingly harmful to islet viability and insulin secretion capabilities. Recent advances in islet cryopreservation technologies have improved outcomes for islet health and survivability during this process. The aim of this chapter is to characterize aspects of the islet cryopreservation method while reviewing current procedural improvements that have led to better outcomes to islet health.

Keywords: cryopreservation, islet, cryoprotectant, alginate, vitrification, diabetes

1. Introduction

Pancreatic islet transplantations are currently used in human clinical studies to treat Type 1 Diabetes Mellitus (T1DM); however, one of the major limitations of this therapy remains efficient and effective storage of islet prior to transplant, during donor-recipient cross-matching [1]. Islet cryopreservation has a distinctly vigorous research history as the storage, transportation, and overall preservation are critical steps in islet transportation. The first human islet allotransplantation took place at the Washington University in the 1980s, which provided proof of insulin independence following the procedure [2]. Following these early successes, clinical trials at the University of Pittsburgh observed further prolonged insulin independence using an improved steroid-free immunosuppressive regimen [3]. These initial clinical islet transplantations demonstrated the need for a preservation process to bridge a temporal gap between islet isolation from donor and transplantation of islet graft into recipients. In 1989, clinical trials at the University of Alberta were able to use both freshly isolated islets supplemented with cryopreserved islets in two T1DM patients, which resulted in partial graft function [4]. The international trial on Edmonton protocol reported success after some of over 43 Type 1 diabetic patients achieved either partial or complete insulin independence for up to 3–5 years post-transplantation [5, 6]. After the establishment of the Edmonton protocol, the Clinical Islet Transplantation Registry (CITR) has recorded more than 1500 human islet allotransplant recipients, which is projected to increase steadily in the future [7]. The Edmonton method of islet isolation was shown to improve islet survival during islet cryopreservation as well. A study comparing cryopreserved islets before the establishment of the Edmonton protocol to human islets treated via Edmonton method observed a 24-hour survival rate increase of 19.3% (50.1 versus 69.4% respectively) with added increases recorded after 7 days of culture [8]. Although the impact of the Edmonton protocol on cryopreservation is significant, there is still vast room for improvement in the islet cryopreservation process.

A major problem of the transplantation field is the lack of human donors' sources; the islet transplantation process is particularly hard-hit from this problem since each transplant recipient must be infused with islets from multiple pancreases [9]. One recipient may also require multiple infusions of donor islets, which can further strain the donor source [10, 11]. With the epidemiologic increase in IDDM diagnosis, islet allotransplantation islet supply will be increasingly strained by the growing demand for islet replacement therapy with projected increases in the population of IDDM individuals [12]. The islet donor supply problems can be partially addressed from improvements in human islet yield, purity, and function. Although an average human cadaveric pancreas contains over 1 million islets, the human islet isolation process can be especially harsh on the isolation yield resulting in loses of up to 50% depending on the degree of success of the isolation [13, 14]. Cryopreservation techniques have been employed to address some of these islet isolation and preservation issues before islets are implanted into the recipient.

The human islet isolation and culturing process involves several steps that vary in temperature, each of which has its own benefits and deficits regarding the health of the islets [13]. Islet procurement from a whole donor pancreas first exposes islet to 4°C (histidine-tryptophan ketoglutarate solution) during sterile transportation to an approved clinical islet center. During the pancreas digestion and purification, the islets are exposed to varying temperatures between 4°C and room temperature. Preservation of islet currently involves the cooling of isolated islets in a temperature-regulated solution at 4°C prior to culturing as prolonged warm-ischemia will increase islet death and subsequent decrease in islet yield [13, 15, 16]. The final step of islet preparation involves a 24- to72-hour culture in approved islet media at 37°C, which has been shown to improve islet yield and functions with reductions in dead/apoptotic islet cell mass [17, 18]. Isolated human islets are preserved while donor islets of similar cross-matching biocompatibility are compiled and matched to an islet recipient [13]. During this critical time period, cryopreservation has been suggested as a preservation method for use during the pre-transplantation period to improve islet health. Improvements in the islet cryopreservation field can translate directly to improvements in human islet cyro-banking as well. This book chapter outlines the history of islet cryopreservation, current techniques in freezing/thawing periods, and cryoprotective additives.

1.1 History of islet cryopreservation

Following the discovery of the microscope, Spallanzani observed that sperm could maintain mobility even when exposed to cold temperature conditions in

1776 [19]. Research into the effects of cryopreservation on live tissue had its roots in late 1800s when scientists used this technology to preserve both spermatozoa and red blood cells (RBCs). During this time, research demonstrated weaknesses in the process which caused inconsistent results and frequent infertility caused by early embryonic death. A breakthrough occurred in the 1950s when James Lovelock discovered that the cryopreservation process caused osmotic stress in the cell by instantly freezing the liquid and causing the formation of ice crystals in RBCs. In 1963, Mazur et al. were able to characterize that process when they demonstrated that the rate of temperature change within a cell-containing medium controlled the movement of water across a cell membrane and thus the degree of intracellular freezing [20]. This together helped to improve the overall understanding of the mechanism associated with the cryoprotective process. During the 1980s, research surrounding the cryopreservation process revealed that the speed at which the freezing and thawing process occurred was the most important factor in determining the survivability of the cells [21, 22]. It was demonstrated that small, slow increments in both the freezing and thawing processes prevented the rapid formation of ice crystals and increased membrane-bound solutes associated with early cell death [23]. Another initial advance in cryopreservation occurred in the late 1940s when researchers discovered that the use of glycerol as a medium increased the survivability of spermatozoa in subfreezing (−70°C) temperatures [24]. Using glycerol as a medium effectively served to protect the cells from rapid formation of ice crystal during the preservation process. A commonly used cyroprotective agent currently employed is dimethyl sulfoxide (DMSO), which is added to cell media prior to the freezing process [25, 26]. DMSO (10%) when added to the cell media, commonly at 2 M concentration, increases the porosity of the cellular membrane, which allows water to flow more freely through the membrane [27, 28]. In addition, early research has demonstrated that nucleation is another way to prevent the rapid formation of ice crystals during freezing [28, 29]. During the freezing process, a metal rod supercooled with liquid nitrogen is applied to the meniscus of the medium containing islets wherein the liquid molecules begin to nucleate. These nucleation reactions are due to the release of latent heat of fusion from the medium, which causes the temperature to decrease more homogenously.

During the 1970s, cryopreservation technology was applied to rat islet preservation in both storage and transportation, which demonstrated maintenance of high viability and function, which showed no significant difference when compared to non-treated islets [30, 31]. In one study, T1DM rats received allogeneic islet transplants cryopreserved using 2 M DMSO (Freezing rate = 0.25°C/min; Thawing rate = 7.5°C), which caused normoglycemia for up to 3 months (**Figure 1**). Additionally, when a modified cryopreservation protocol was applied to canine allotransplantations (freezing rate = 0.25°C/min; thawing rate = 3.4°C/min), T1DM canine recipients demonstrated prolonged glycemic control for up to 18 months [32]. These results highlighted the potential use of cryopreservation for islet transplantations in both small and large animal models in addition to differences required when cryofreezing small and large animal pancreases. Recent advances in islet quality control like oxygen consumption rate (OCR), qPCR, and MTT [3-(4,5-dimethylthiazol-2-yl)-2,5-diphenyltetrazolium bromide] assay allow researchers to study islet cryopreservation technology more effectively [33]. Islet cryopreservation research is a very active research topic with many studies aiming to characterize and improve cryopreservation freezing/thawing processing, benefits of potential cyroprotective additives, and the effects of encapsulation on islet function during cryopreservation. The following section will discuss the islet cryopreservation process (**Figure 2**).

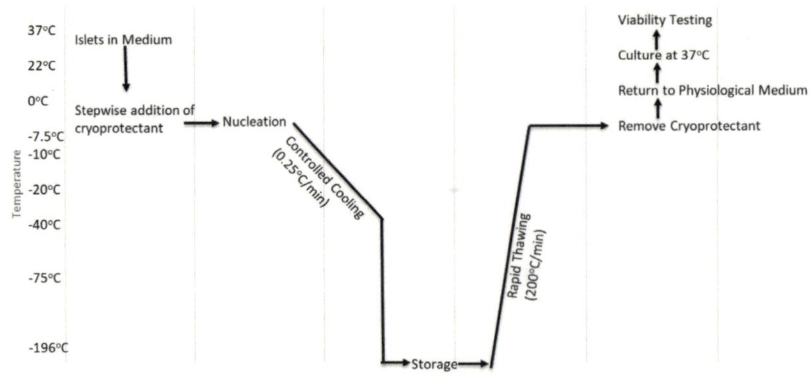

Figure 1.
Sample image of encapsulated human islets stained with dithizone for 15 min, taken at 2× magnification with objective lens 20/40 PH. Scale bar represents 2 mm. Imaging performed at UCI laboratory under supervision of Dr. Jonathan Lakey PhD.

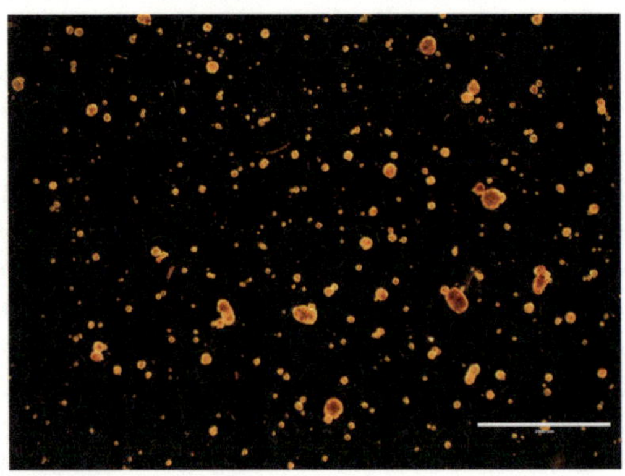

Figure 2.
Flowchart of cryopreservation. This chart describes the range of temperature, rate of temperature change, and the procedure involved during cryopreservation.

2. Characteristics of the islet cryopreservation process

2.1 Background

A crucial aspect of cryofreezing islets is the rate of freezing and thawing, which can have major effects on the islet health and morphology. The freezing process describes the process of cooling the islet-containing medium to around −196°C. If the freezing process is done too rapidly (>0.25°C/min), the liquid in the medium will freeze too quickly and crystal ice structure will form within cell membranes. Conversely, if the freezing process is performed too slowly (<0.1°C/min), then innate/adaptive immune cells, such as macrophages, dendritic cells, and lymphocytes which are present within the islet medium survive in greater numbers and can contribute to foreign body response (FBR)-mediated graft rejection upon transplantation [3, 34]. Taylor et al. demonstrated increase in macrophage viability (91%) cryofreezing is done at a rate of 0.1°C/min compared to 72–75% viable macrophage when the rate between 0.1–20°C/min [35]. Therefore, a key aspect of cryofreezing is the use of an optimal freezing rate based on islet type and volume to prevent ice crystal formation and immune cell survival. Over the years, many studies have described varying optimal freezing rates, which has made it difficult to compare between freezing/thawing methods. A consistent freezing/thawing protocol uses slow freezing from −40 to −196°C followed by a rapid thawing starting from −196°C [36]. An early study aimed at characterizing differences between cooling and thawing rates exposed islets to several freezing rates between 0.3 and 100°C/min and thawing rates of 10 or 50°C/min. Highest survival rates were detected at 0.3°C/min rate with slight decreases observed between 60 and 1000°C/min rates [37]. The study also demonstrated the critical nature of using DMSO as a medium to protect the islet viability and function from cooling process. Cyroprotective agents (CPAs) like DMSO are neutral solutes of both low toxicity and molecular weight that replace up to 30% of the cell water and provide optimal conditions for subzero temperatures [38]. While a variety of DMSO concentrations have been tested, the most popular one used is 2 M DMSO, which is added in a stepwise fashion (1 M DMSO to 2 M DMSO) during pre-freezing [36]. One study found that when islets were exposed to 1 M DMSO for 30 min followed by incubation in 2 M DMSO for 10 min before cooling phase, then the islet insulin secretory patterns were improved after thawing [39]. More recently, studies have shown that the rate of cooling is much less important than the use of cryoprotective additives during the pre-freezing phase [36, 39].

2.2 Vitrification

The process of exposure and equilibration of permeating cyroprotective additives to islets is known as vitrification, which was first described by Rall et al. [38, 40]. While the use of CPAs reduces the risk of rapid ice crystal formation during the cooling phase, cryoprotective agents, such as DMSO and ethylene glycol, have been shown to be toxic to islet viability and function when concentrated in the medium [38, 41, 42]. Vitrification is used to slow the exposure of islets to CPAs by adding the CPAs in a stepwise fashion, usually in ascending concentrations of CPA, thereby allowing the CPAs to slowly permeate and form a solute equilibrium across the cell membrane. The vitrification process also causes water to flow extracellularly where, during cooling, vitreous water crystals slowly form outside the islet cells [38]. This vitreous medium exists in a solid-liquid transition state that is maintained at a supercooled temperature (≤100°C), thus having the structure of a liquid but behaving mechanically like a solid [43]. In addition to reducing ice crystal formation, vitrification involves exposing the CPAs in a stepwise fashion (1 M DMSO followed by 2 M DMSO), thus reducing the toxicity of the CPAs as the cooling process proceeds. Once the cooling process is finished, the cryopreserved islets will be stored at −196°C liquid nitrogen freezer until use.

2.3 Thawing

When the supercooled islet is in −196°C storage, the vitreous medium is still locked in a liquid-solid transition state; however, once the warming process begins, there is risk of ice recrystallization within the medium, which can damage the islets [36, 43]. A widely accepted procedure includes the use of rapid thawing from −196°C at a rate of 150–200°C/min. Mechanical agitation is applied to thawing samples in a 37°C water bath until all ice is gone (0°C). At 0°C, the samples are then placed in an ice slush bath (0°C) until thawing is completed [36]. Following the warming phase, it is important to remove the cryoprotectant from the medium. After thawing is complete, samples are spun down at 1500 RPM and supernatants containing most of the cryoprotectants are removed. The pellet is then resuspended in an isotonic buffer (0.75 M sucrose) to dilute out the remaining CPAs.

3. Factors that affect islet cryopreservation process

3.1 Background

While the use of CPAs is a critical step, there are several factors that play a significant role in the success of an islet cryopreservation process. Technological advances in these areas can help reduce the stress of the process on the islets while more effectively processing the islets tissue during cryopreservation.

3.1.1 Oxygen treatment

Changes in ambient atmospheric oxygen concentration can have a negative effect on islet viability and function via ATP metabolite depletion. In hypoxic conditions, β-cells are put under oxidative stress followed by ROS production, which facilitates islet death [44]. The thawing and rewarming phases have been shown to place hypoxic-related oxidative stress on cryopreserved tissues [45]. During the thawing/rewarming period, as islets are brought back from sub-freezing temperature, cellular enzymes begin to function and increase oxygen consumption. This process can cause a reduction in adenosine 3-phosphate (ATP), a cellular metabolite, which can lead to islet death [45]. To address this issue, researchers hypothesized that hyperbaric conditions might improve islet recovery. Human islets were exposed to both normal oxygen conditions (21% O_2, 74% N_2, and 5% CO_2) and hyperbaric conditions (50% O_2, 45% N_2, and 5% CO_2) at 22°C for 45 min followed by 37°C for 45 min. Short-term (post-rewarming) and long-term (2-day culture) islet function analysis was conducted via GSIS, qPCR ischemia-gene analysis, and islet metabolism via oxygen consumption rate (OCR) assay. Long-term culture also compared normal and hyperbaric culture conditions. No significant short-term function and metabolic differences were observed between conditions and non-treated islets. However, the hyperbaric conditions were shown to suppress the increases of inflammation detected in untreated cryopreserved islets. Islet recovery after long-term culture was significantly better under hyperbaric conditions and was shown to increase after hyperbaric 2-day culture. A recent advancement in organ oxygenation known as persufflation involves the perfusion of humidified oxygen into the vasculature of the pancreas before and after cryopreservation. This technique could potentially mitigate the islet loss from hypoxia and ischemia during cryofreezing, thawing, and rewarming phase.

3.1.2 Cryopreservation storage duration

Development of an islet tissue bank will require islets to be cryopreserved for long periods of time, years if necessary. Generally, islet cryopreservation studies will only cryopreserve islets for short periods of time (1–90 days) [45–47]. For human islet cryopreservation, the upper limit of storage duration was set by Fox et al. in 2015. Human islets were cryopreserved at −196°C for an average of 17.6 ± 0.4 years. Between 2012 and 2014, human islets were then thawed and warmed, after which islet electrophysiology and function were analyzed. After measurement of β-cell excitability via path-clamp assay, similar Ca^{2+}-influx conductance patterns were observed between cryopreserved and fresh islets [48]. However, insulin stimulation index was significantly lower for cryopreserved (1.90 ± 0.24) compared to fresh islets (9.53 ± 0.92). However, after cryopreserved islets were transplanted into STZ-induced mice, partial normoglycemia was observed for 60 days with improvements to glucose tolerance [48]. More research is needed to infer a limit to cryopreservation storage duration to ensure that islets remain functional post-thawing.

3.1.3 Islet structure limitations

Difficulties with the cryopreservation of whole organs are partially due to the non-homogenous temperature distribution within the large tissue structure of organs. Pancreatic islets exist as a spherical cluster, with an average diameter of 100 μm, of several thousands of cells connected by a dense network of connective tissue [49]. The cells that lay within the islet core are susceptible to hypoxia-related stress, particularly during the cryopreservation process [50, 51]. A recent study aimed to address this structural problem by reducing islets to single cells followed by cryopreservation [52]. Islets were reduced to single cells and cryopreserved with 10% DMSO and stored at −196°C (1°C/min) for four or more weeks. After a rapid thawing and warming phase, islets were reaggregated at 37°C [52]. Reaggregated islets were recovered at a rate of 80% and had similar diameter to intact cryopreserved islets. The viability of reaggregated islets was significantly higher than intact islets (80 versus 25% respectively) post-thawing. No significant differences in GSIS function were detected between reaggregated islets and intact islets. Upon allotransplantation of reaggregated islets into omentum of STZ-induced diabetic rats, normoglycemia was achieved in 24-hours and was sustained for 10-months. Intact cryopreserved islets failed to achieve normoglycemia. Graft volume necessary to achieve diabetic correction was lower for reaggregated islets (5–8500 IE/kg) than fresh islets (10–12,000 IE/kg) [52].

4. Advances in islet cryoprotective additive technology

4.1 Principles background

CPAs can be divided into two major types, namely permeating and non-permeating additives [53]. The main difference between the two major sub-types is whether the substance can penetrate the intracellular space during vitrification [54, 55]. Since the accidental discovery of glycerol in the 1940s [24], penetrating CPAs, such as DMSO and ethylene glycol, have shown significant benefits for islet survival in many studies [36]. These penetrating cryoprotectants, usually low-molecular weight polar aprotic solvents, penetrate the cell membrane and increase the inner volume of the cell. An equilibrium is reached across the cell membrane when the intracellular water content is lower than physiological normal range, thus reducing the probability of intracellular ice crystal formation [56, 57]. Non-penetrating cryoprotective additives like saccharides, which have a large molecular weight, remain in the extracellular space during the freezing process [56]. A build-up of these molecules in the extracellular space induces an osmotic gradient across the cell membrane, which causes water to move out of the cell. Water movement into the extracellular space helps to reduce the risk of intracellular ice crystal formation in addition to depressing the freezing point of intracellular water [58–60]. The mechanism of non-penetrating CPAs has been demonstrated to be temperature sensitive and suboptimal for certain cell types; therefore, penetrating CPAs have been traditionally favored over non-penetrating CPAs even though penetrating CPAs have increased the risk of toxicity [61, 62]. Recently, the integration of both permeating and non-permeating CPAs (e.g., DMSO with University of Wisconsin Solution (UW)) has shown improvements to post-cryopreservation islet recovery and insulin secretory behavior of islets.

4.2 Permeating cryoprotective additives

4.2.1 Dimethyl sulfoxide (DMSO)

DMSO toxicity toward islets has been shown to be minimal at concentrations used during the freezing phase and has even demonstrated protective capabilities against selective β-cell necrosis antagonist alloxan [38, 41, 63]. DMSO is considered the gold standard in islet cryoprotective additives and has been heavily used in research for the prevention of intracellular ice crystal formation. A 1999 study sought to compare the effect of DMSO-mediated cryopreservation on the recovery and function of canine islets [64]. Islets from seven consecutive canine isolations were dissociated into single cells and cryopreserved in 2 M DMSO medium using a slow stepwise cooling method (0.25°C/min) to 40°C followed by storage in −196°C. Following rapid thawing (200°C/min), 81.5% of cryopreserved islets were recovered with no significant difference in insulin stimulation index (SI) when compared to non-treated canine islets (10.5 A.U. versus 12.4 A.U. respectively) [64]. Another study sought to standardize the critical removal process of DMSO from islet medium during the thawing phase. This protocol involves the slow stepwise addition of sucrose solution to dilute out the DMSO post-thawing [64]. Overall, DMSO will continue to play an important role in islet cryopreservation research.

4.2.2 Ethylene/polyethylene glycol

A common constituent of car antifreeze, other permeating CPAs include both ethylene and polyethylene glycol, which have been studied for islets cryopreservation [53]. These low-molecular weight substances easily penetrate the cell membrane, much like DMSO, and cause solute equilibrium, which osmotically drives water toward the extracellular space [38]. Once the use of DMSO as a CPA was established in islet cryofreezing, studies in rat islets began to suggest potential toxicity issues when DMSO was exposed to rat islets [41, 65]. One study comparing DMSO and EG CPAs resulted in DMSO islets that exhibited lower cellular DNA, insulin, glucagon, and impaired insulin secretory patterns compared to EG, which was more like non-frozen islets. Upon transplantation of each islet group, normal glycemic control was achieved in 100% of EG-treated and non-frozen islets but only 92% of DMSO-treated islets recipients, which also experienced delays in diabetes correction [65]. When islets cryopreserved with varying concentrations (1, 2, and 3 M) of DMSO, EG, and PG were exposed to islets from canine and human sources, the permeability (P_s = μm/s) was quantified. The highest P_s was achieved in canine islets when 2 M EG (2.47 μm/s) was used while 2 M PG showed the highest P_s in human islets (3.48 μm/s) suggesting potential use of EG and PG in islet cryopreservation [66].

4.2.3 Permeating CPA mixtures

Attempts have been made to produce mixtures of DMSO and EP (30% EP, 20% DMSO) for use during vitrification phase, which can help reduce the toxicity risks of using DMSO alone. A mixture of ethylene glycol (EG) and DMSO, classified as EDT324, was used as a cryoprotectant during the cooling phase with rat islets. EDT324-treated cryopreserved islets showed significant increases in islet viability and insulin secretory capability compared to use of DMSO (10%) alone [67]. EDT324-treated islets were then transplanted into allogenic rat recipients and diabetic correction was achieved after 2 days. Similar results were observed after islets were treated with one of two EG/DMSO mixtures (1 M ME2SO + 1 M EG, or 1 M ME2SO + 0.5 M EG). Islets treated with permeating CPA mixtures achieved significantly higher yield and viability compared to islets treated with DMSO only. When transplanted into STZ-induced mice, islets treated with DMSO/EG mixtures caused normoglycemia 12 days faster on average than DMSO only-treated islets [68].

4.3 Non-permeating cryoprotective additives

4.3.1 Saccharides

Although permeating cryoprotectants have been mainly used during mammalian cell cryopreservation, saccharides have demonstrated survival advantages when added to the vitrification medium. When adult human islets were treated with 300 mmol/L trehalose, a 92% recovery rate was achieved compared to 58% recovery of DMSO-treated islets in addition to 14-fold increase in insulin content within islet grafts. More prominent differences in recovery were observed in fetal human islets treated with trehalose compared to islet only treated with DMSO [69]. More recently, an antifreeze glycoprotein (AFGP) was included to DMSO slow cooling phase medium during cryofreezing of rat islets. When compared to DMSO only protocol, AFGP-treated islets demonstrated significant increases in recovery rate (85 ± 6.25 versus 63.3 ± 14.2%) and insulin stimulation index (3.86 ± 0.43 versus 2.98 ± 0.22) were observed [70]. These results demonstrate that saccharides and saccharide-containing substances can be used in conjunction with lower DMSO concentrations and help reduce islet toxicity.

4.3.2 Polymeric compounds

High-molecular weight polymeric compounds such as polyvinylpyrrolidone (PVP) and dextran have been shown to be effective at formation of amorphous glass matrix during cryofreezing phase [71]. When 10% PVP was added to cryopreservation medium before cooling phase, rat islet recovery and function were significantly higher than when islets were treated with 2 M DMSO and 3 M glycerol. Islets treated with 2 M PG demonstrated comparable islet recovery and function to PVP-treated group [72]. Although, the use of high-molecular weight cryoprotectants has been observed, past studies suggest that these compounds are ineffective at slow cooling temperature transitions, which is a crucial step of the cryofreezing process [71].

4.4 Other potential CPAs

One of the most important causes of cell damage/death during cryopreservation is due to ice crystal formation. However, there are unavoidable damaging consequences to islet health when islet cells, especially when in multicellular tissues fragments, are exposed to subfreezing temperatures (≥100°C), which can lead to apoptosis and/or necrosis after the post-thawing phase of cryopreservation [73]. Due to its fragile multicellular tissue structure, islet fragments are susceptible to various stresses including oxidative stress, osmotic stress, hypoxia, hypothermia, and inflammation induced by the cryopreservation process, which can have acute and/or long-term effects on islet graft viability and function [74, 75]. As research into islet cryoprotection has become more nuanced in recent years, studies have started to target CPAs, which reduce stress-induced cell death associated with the cryopreservation process (**Table 1**).

Parameter	Method	Reference
Cryoprotectant	EPA + DHA + Metformin	[40]
Cooling rate	Rapid (50–70°C/min)	[27–29]
Thawing rate	Rapid (150–200°C/min)	[32]
Oxygen environment	50% during thawing	[54]
3D structure	Freeze as individual cells, re-aggregate into spheroids after thaw	[51]
Encapsulation	1.75% Alginate encapsulation prior to cryopreservation	[64]
Assessment methods include islet viability, glucose sensitivity, and GSIS values after thawing.		

Table 1.
Cryoprotection: A quick summary of the parameters that had the best outcome for the islets during cryopreservation.

4.4.1 Butylated hydroxyanisole (BHA)

Oxidative stress in cells produces endogenous reactive oxygen species (ROS), such as superoxide (O_2^-) and free hydroxyl (OH^-), which leads to an increase in free radical concentration intracellularly [76]. Elevated internal levels of free radicals can cause cellular damage and lead to cellular process disruptions. To combat cryopreservation-related oxidative stress, one early study added butylated hydroxyanisole (BHA) to islet cryomedium while monitoring oxidative stress via glutathione redox state (GSH/GSSG). Islets treated with BHA demonstrated enhanced insulin secretory behavior (2.2-fold increase) when compared to untreated islets. In addition, exposure to alloxan, a highly damaging free radical generating agent, did not induce significant oxidative stress [77].

4.4.2 Ascorbic acid-2 glucoside (AA2G)

Ascorbic acid-2 glucoside (AA2G), a derivative of Vitamin C, is a potent antioxidant and can deliver stable antioxidant activity into culture media [78]. AA2G (100 μg/mL) in combination with the UW islet preservation solution was used as the cryopreservation medium [79, 80]. Following 3 months of storage (−80°C), the islets treated with UW/AA2G demonstrated viability maintenance (68.3 ± 5.6%) and significantly increased insulin stimulation via glucose-stimulated insulin secretion (GSIS) test, when compared to treatment with UW alone (1.93 ± 0.5 and 1.17 ± 0.6 respectively). Transplantation of thawed AAG2/UW-treated into liver of nude mice produced engraftment with insulin-positive cells observed.

4.4.3 Curcumin

Curcumin, the main component of turmeric spice, has demonstrated antioxidant and anti-inflammatory effects in multiple cell types [81]. Curcumin has not been shown to increase insulin stimulation; however, it has demonstrated upregulation of oxidative stress-reducing genes Hsp70 and HO-1 [47]. To evaluate cyroprotective abilities of curcumin, Kanitkar et al. compared the effect of 10% DMSO with and without 10 μM curcumin on islets treated with slow cooled cryopreservation (−196°C) for 7 days. Curcumin-treated islets showed increases in SI compared to non-treated cryopreserved islets but no difference from fresh islets.

In curcumin-treated medium, over-expression of HO-1 and Hsp70 was observed to incrementally increase as the cryopreservation process unfolded [47].

4.4.4 Taurine

An amino sulfonic acid, taurine has been suggested as a CPA because of chemical properties that allow antioxidant and osmoregulatory properties [82]. An addition of taurine to the cryopreservation medium demonstrated cyroprotective effects during islet cryopreservation (Hardikar 2001). Pretreatment of taurine prior to cryopreservation freezing phase at 0.3 mM and 3.0 mM resulted in high maintained viability of 91.9 ± 2.3 and 94.6 ± 1.58% respectively. Lipid peroxidation, which is a known indicator of oxidative stress, was reduced significantly compared to controls. Finally, normoglycemia was achieved when taurine-treated cryopreserved islets were transplanted into immunocompetent mice.

4.4.5 Γ-aminobutyric acid (GABA)

GABA neurotransmitters facilitate the inhibitory neuronal pathways within the central nervous system and have demonstrated regulatory and protective effects on β-cells [83]. GABA has been shown to produce membrane polarization (Ca^{2+}-influx) which activates survival and growth pathways (PI_3-K/Akt) and can restore β-cells' mass in severely diabetic mice [83, 84]. Islets were treated with 50 or 100 μM GABA and were cryopreserved to 196°C for up to 30 days. At both 15 and 30 days post-thawing, islets treated with GABA exhibited similar insulin secretory behavior compared to fresh islets. When oxidative stress was measured via MTT assay, reduced ROS content was observed in GABA-treated islets in comparison to non-treated cryopreserved islets [46].

4.4.6 Metformin

Metformin is a standard-of-care drug used for the treatment of Type 2 Diabetes and has been identified as an "essential medicine" by the WHO [85]. While the complete mechanism of action is unknown for metformin, it has demonstrated insulin-sensitizing properties and reduces unfettered liver gluconeogenesis [86, 87]. When used as a cryoprotective agent at ultralow temperatures, metformin produced membrane stabilizing effects. Cryopreserved islets treated with metformin-containing cyromixtures exhibited comparable viability (90 versus 100%) to non-treated fresh islets. Improved insulin secretion was observed at 15 days post-thawing (8 ng/mL) with a stimulation index value of >5, suggesting islets were highly functional [46]. Recently, similar cyroprotective results were observed in chicken islets [88].

4.4.7 Sericin

Produced by *Bombyx mori* silkworm, this gel-like protein has previously demonstrated oxidative stress reduction properties induced by freezing temperatures in both rat islets and various other mammalian cell lines [89, 90]. When added to media, Ahnishi et al. showed no significant differences between GSIS results between the FBS + DMSO and the sericin + DMSO groups. This study demonstrated that reduction in DMSO content in cyromixtures is possible with addition of 1% sericin, which could reduce the toxic risk posed by high concentrations of DMSO.

4.5 Recent advances in islet cryopreservation technology

4.5.1 Cryopreservation with alginate-based microencapsulation technology

Alginate-based microencapsulation technologies have developed in concert with islet transplantation research where the alginate polymer forms a semipermeable immune-isolating barrier around islet fragments [91–93]. Alginate-microencapsulation has recently been applied to the field of islet cryopreservation in order to characterize its effect on islet survivability [33]. Chen et al. reported the development of an oxygen-sensing alginate coating to encapsulate islets prior to cryopreservation [94, 95]. Islets were encapsulated with alginate coating containing ruthenium-based oxygen-sensitive fluorophore (ROF) after which the encapsulated islets were subjected to a 10% DMSO with or without $50x10^{-3}$ M trehalose and stored for 1–7 days. Encapsulated islets undergoing cryopreservation showed significantly higher insulin stimulation behavior than bare islets at Day 1 and 7 [94]. In addition to cyroprotective abilities, the microcapsules treated with ROS demonstrated viable oxygen sensitivity during OCR measurements. This study demonstrates the use of multiple cyroprotective parameters to mitigate potential damage during cryopreservation [94]. Other studies have demonstrated the benefits of alginate microencapsulation use during cryopreservation as the 3D barrier is porous and can resist stress/strain associated with ice formation [94, 96, 97].

4.5.2 Hollow fiber vitrification

Previously described for use in embryological studies, vitrification scaffolds have been suggested as a medium for cryopreservation of islets to improve the islet survival [98, 99]. This technique involves the loading of islets into a hollow fiber chamber (HFV) composed of cellulose-triacetate, which is permeable to CPAs [100]. Researchers used a combination of permeating CPAs like DMSO and EG during cooling phase of both non-vitrified islets and islets vitrified using HFV method. *In Vitro* assays demonstrated similar islet structure and insulin gene promoter expression (NeuroD, Pdx1, MafA) to non-vitrified islets; however, the insulin stimulation index was significantly decreased for islets undergoing HFV compared to non-vitrified islets (27.8 ± 8.2 and 3.5 ± 0.6 respectively). Nonetheless, after HVF and non-vitrified islets were transplanted into kidney subcapsular space, all mice were euglycemic within 4–8 days and remained so for 1 month until nephrectomy, which induced hyperglycemia.

5. Conclusion

Islet cryopreservation has come a long way in the last 40 years. Many parameters of cryopreserving islets are being actively researched because of the high demand for long-term storage. Currently, entire organ cryopreservation is not entirely feasible, and is only shown possible after a few hours of storage [101]. Based on experiments performed in our lab along with results from other research groups, we predict that future improvements in islet cryopreservation will rely on the use of a mixture of cyroprotective additives along with the use of secondary technologies like alginate encapsulation [96]. We hope that with a standardized cryopreservation protocol, islet banking would be more feasible, and ultimately, transplantation would no longer be throttled by the donor-recipient mismatch.

Acknowledgements

The authors gratefully acknowledge the support from the Department of Surgery and the Department of Biomedical Engineering at University of California, Irvine, and the Sue and Bill Gross Stem Cell Research Center, for all the support in the writing of this chapter.

Author details

Samuel Rodriguez[1], David Whaley[1], Michael Alexander[1], Mohammad Rezaa Mohammadi[2] and Jonathan R.T. Lakey[1,3]*

1 Department of Surgery, University of California Irvine, Orange, CA, USA

2 Department of Materials Science and Engineering, University of California Irvine, Irvine, CA, USA

3 Department of Biomedical Engineering, University of California Irvine, Irvine, CA, USA

*Address all correspondence to: jlakey@uci.edu

References

[1] Bruni A, Gala-Lopez B, Pepper AR, Abualhassan NS, Shapiro AJ. Islet cell transplantation for the treatment of type 1 diabetes: Recent advances and future challenges. Diabetes, Metabolic Syndrome and Obesity: Targets and Therapy. 2014;7:211-223

[2] Scharp DW, Lacy PE, Santiago JV, et al. Insulin Independence after islet transplantation into type I diabetic patient. Diabetes. 1990;**39**(4):515-518

[3] Ricordi C, Tzakis AG, Carroll PB, et al. Human islet isolation and allotransplantation in 22 consecutive cases. Transplantation. 1992;**53**(2):407-414

[4] Warnock GL, Kneteman NM, Ryan EA, et al. Continued function of pancreatic islets after transplantation in type I diabetes. The Lancet. 1989;**334**(8662):570-572

[5] Shapiro AMJ, Ricordi C, Hering BJ, et al. International trial of the Edmonton protocol for islet transplantation. New England Journal of Medicine. 2006;**355**(13):1318-1330

[6] Shapiro AMJ, Lakey JRT, Ryan EA, et al. Islet transplantation in seven patients with type 1 diabetes mellitus using a glucocorticoid-free immunosuppressive regimen. New England Journal of Medicine. 2000;**343**(4):230-238

[7] Registry CIT. Scientific Summary of the Collaborative Islet Transplant Registry (CITR) 2015 (Tenth) Annual Report. 2015; Available from: https://citregistry.org/system/files/10AR_Scientific_Summary.pdf

[8] Miranda PM, Mohan V, Ganthimathy S, et al. Human islet mass, morphology, and survival after cryopreservation using the Edmonton protocol. Islets. 2013;**5**(5):188-195

[9] Park C-G, Bottino R, Hawthorne WJ. Current status of islet xenotransplantation. International Journal of Surgery. 2015;**23**:261-266

[10] Organ Procurement and Transplantation Network. 2019. Available from: https://optn.transplant.hrsa.gov/data/organ-datasource/pancreas/ [Accessed: 4-11-2019]

[11] Ekser B, Cooper DKC, Tector AJ. The need for xenotransplantation as a source of organs and cells for clinical transplantation. International Journal of Surgery. 2015;**23**:199-204

[12] Soto-Gutierrez A, Wertheim JA, Ott HC, Gilbert TW. Perspectives on whole-organ assembly: Moving toward transplantation on demand. The Journal of Clinical Investigation. 2012;**122**(11):3817-3823

[13] Shapiro AMJ, Pokrywczynska M, Ricordi C. Clinical pancreatic islet transplantation. Nature Reviews Endocrinology. 2016;**13**:268

[14] Benhamou PY, Watt PC, Mullen Y, et al. Human islet isolation in 104 consecutive cases. Factors affecting isolation success. Transplantation. 1994;**57**(12):1804-1810

[15] Tanioka Y, Hering BJ, Sutherland DER, et al. Effect of pancreatic warm ischemia on islet yield and viability IN DOGS1. Transplantation. 1997;**64**(12):1637-1641

[16] Corlett MP, Scharp DW. The effect of pancreatic warm ischemia on islet isolation in rats and dogs. Journal of Surgical Research. 1988;**45**(6):531-536

[17] Murdoch TB, McGhee-Wilson D, Shapiro AMJ, Lakey JRT. Methods of human islet culture for transplantation. Cell Transplantation. 2004;**13**(6):605-618

[18] Gaber AO, Fraga DW, Callicutt CS, Gerling IC, Sabek OM, Kotb MY. Improved IN vivo pancreatic islet function after prolonged IN vitro islet culture. Transplantation. 2001;**72**(11):1730-1736

[19] Walters EM, Benson JD, Woods EJ, Critser JK. The history of sperm cryopreservation. In: Sperm Banking Theory and Practice. London: Cambridge University Press; 2009. pp. 1-17

[20] Mazur P. Kinetics of water loss from cells at subzero temperatures and the likelihood of intracellular freezing. The Journal of General Physiology. 1963;**47**(2):347-369

[21] Chesné C, Guillouzo A. Cryopreservation of isolated rat hepatocytes: A critical evaluation of freezing and thawing conditions. Cryobiology. 1988;**25**(4):323-330

[22] Al-Hasani S, Diedrich K, Van der Ven H, Reinecke A, Hartje M, Krebs D. Cryopreservation of human oocytes. Human Reproduction. 1987;**2**(8):695-700

[23] Mazur P. Freezing of living cells: Mechanisms and implications. American Journal of Physiology-Cell Physiology. 1984;**247**(3):C125-C142

[24] Polge C, Smith AU, Parkes AS. Revival of spermatozoa after vitrification and dehydration at low temperatures. Nature. 1949;**164**(4172):666

[25] Siebzehnruebl E, Todorow S, Van Uem J, Koch R, Wildt L, Lang N. Cryopreservation of human and rabbit oocytes and one-cell embryos: A comparison of DMSO and propanediol. Human Reproduction. 1989;**4**(3):312-317

[26] Friedler S, Giudice LC, Lamb EJ. Cryopreservation of embryos and ova. Fertility and Sterility. 1988;**49**(5):743-764

[27] Ock S-A, Rho G-J. Effect of dimethyl sulfoxide (DMSO) on cryopreservation of porcine mesenchymal stem cells (pMSCs). Cell Transplantation. 2011;**20**(8):1231-1239

[28] Bouquet M, Selva J, Auroux M. Effects of cooling and equilibration in DMSO, and cryopreservation of mouse oocytes, on the rates of in vitro fertilization, development, and chromosomal abnormalities. Molecular Reproduction and Development. 1995;**40**(1):110-115

[29] Paynter SJ, Fuller BJ. Cryopreservation of mammalian oocytes. In: Cryopreservation and Freeze-Drying Protocols. Switzerland: Springer; 2007. pp. 313-324

[30] Andersson A, Sandler S. Viability tests of cryopreserved endocrine pancreatic cells. Cryobiology. 1983;**20**(2):161-168

[31] Bretzel R, Schneider J, Dobroschke J, Schwemmle K, Pfeiffer E-F, Federlin K. Islet transplantation in experimental diabetes of the rat VII. Cryopreservation of rat and human islets. Preliminary results. Hormone and Metabolic Research. 1980;**12**(06):274-275

[32] Rajotte RV, Warnock GL, Kneteman NN. Cryopreservation of insulin-producing tissue in rats and dogs. World Journal of Surgery. 1984;**8**(2):179-186

[33] Benson E, Harding K, Ryan M, Petrenko A, Petrenko Y, Fuller B. Alginate encapsulation to enhance biopreservation scope and success: A multidisciplinary review of current ideas and applications in cryopreservation and non-freezing storage. CryoLetters. 2018;**39**(1):14-38

[34] Rajotte R, Warnock G, Kneteman N, Erickson C, Ellis D. Optimizing cryopreservation of isolated islets. Paper presented at: Transplantation proceedings: 1989

[35] Taylor M, Bank H. Function of lymphocytes and macrophages after cryopreservation by procedures for pancreatic islets: Potential for reducing tissue immunogenicity. Cryobiology. 1988;**25**(1):1-17

[36] Rajotte RV. Islet cryopreservation protocols. Annals of the New York Academy of Sciences. 1999;**875**(1):200-207

[37] Taylor MJ, Benton MJ. Interaction of cooling rate, warming rate, and extent of permeation of cryoprotectant in determining survival of isolated rat islets of Langerhans during cryopreservation. Diabetes. 1987;**36**(1):59-65

[38] Taylor MJ, Baicu S. Review of vitreous islet cryopreservation: Some practical issues and their resolution. Organogenesis. 2009;**5**(3):155-166

[39] Foreman J, Moriya H, Taylor M. Effect of cooling rate and its interaction with pre-freeze and post-thaw tissue culture on the in vitro and in vivo function of cryopreserved pancreatic islets. Transplant International. 1993;**6**(4):191-200

[40] Rall WF, Fahy GM. Ice-free cryopreservation of mouse embryos at – 196 C by vitrification. Nature. 1985;**313**(6003):573

[41] Sakonju I, Taura Y, Inayoshi Y, et al. Cryopreservation of isolated rat islets of Langerhans in the presence of ethylene glycol or dimethyl Sulfoxide: Evaluation of toxicity and the dynamic pattern of subsequent insulin Releasein vitro. Cryobiology. 1996;**33**(3):354-362

[42] Fahy GM. The relevance of cryoprotectant "toxicity" to cryobiology. Cryobiology. 1986;**23**(1):1-13

[43] Debenedetti PG, Stillinger FH. Supercooled liquids and the glass transition. Nature. 2001;**410**(6825):259

[44] Xu J, Long Y-S, Gozal D, Epstein PN. β-Cell death and proliferation after intermittent hypoxia: Role of oxidative stress. Free Radical Biology and Medicine. 2009;**46**(6):783-790

[45] Vajdová Kn SR, Mišlanová C, Kukan M, Lutterová M. Cold-preservation–induced sensitivity of rat hepatocyte function to rewarming injury and its prevention by short-term reperfusion. Hepatology. 2000;**32**(2):289-296

[46] Chandravanshi B, Dhanushkodi A, Bhonde R. High recovery of functional islets stored at low and ultralow temperatures. The Review of Diabetic Studies : RDS. 2014;**11**(3):267

[47] Kanitkar M, Bhonde RR. Curcumin treatment enhances islet recovery by induction of heat shock response proteins, Hsp70 and heme oxygenase-1, during cryopreservation. Life Sciences. 2008;**82**(3-4):182-189

[48] Fox JEM, Lyon J, Dai XQ, et al. Human islet function following 20 years of cryogenic biobanking. Diabetologia. 2015;**58**(7):1503-1512

[49] Brissova M, Fowler MJ, Nicholson WE, et al. Assessment of human pancreatic islet architecture and composition by laser scanning confocal microscopy. The Journal of Histochemistry and Cytochemistry. 2005;**53**(9):1087-1097

[50] Sakata N, Chan NK, Ostrowski RP, et al. Hyperbaric oxygen therapy improves early posttransplant islet function. Pediatric Diabetes. 2010;**11**(7):471-478

[51] Lehmann R, Zuellig RA, Kugelmeier P, et al. Superiority of small islets in human islet transplantation. Diabetes. 2007;**56**(3):594-603

[52] Rawal S, Harrington S, Williams SJ, Ramachandran K, Stehno-Bittel L. Long-term cryopreservation of reaggregated pancreatic islets resulting in successful transplantation in rats. Cryobiology. 2017;**76**:41-50

[53] Karow AM Jr. Cryoprotectants—A new class of drugs. Journal of Pharmacy and Pharmacology. 1969;**21**(4):209-223

[54] McGann LE. Differing actions of penetrating and nonpenetrating cryoprotective agents. Cryobiology. 1978;**15**(4):382-390

[55] Taylor R, Adams G, Boardman C, Wallis R. Cryoprotection—Permeant vs nonpermeant additives. Cryobiology. 1974;**11**(5):430-438

[56] Pegg DE. Principles of cryopreservation. In: Preservation of Human Oocytes. Florida, USA: CRC Press; 2009. pp. 34-46

[57] Lyman GH, Preisler HD, Papahadjopoulos D. Membrane action of DMSO and other chemical inducers of friend leukaemic cell differentiation. Nature. 1976;**262**(5567):360

[58] Bhatnagar BS, Cardon S, Pikal MJ, Bogner RH. Reliable determination of freeze-concentration using DSC. Thermochimica Acta. 2005;**425**(1-2):149-163

[59] Takahashi T, Hirsh A, Erbe E, Williams RJ. Mechanism of cryoprotection by extracellular polymeric solutes. Biophysical Journal. 1988;**54**(3):509-518

[60] Doebbler G, Sakaida R, Cowley C, Rinfret A. Cryogenic preservation of whole blood for transfusion in vitro study of a process using rapid freezing, thawing and protection by polyvinylpyrrolidone. Transfusion. 1966;**6**(2):104-111

[61] Newton H, Fisher J, Arnold J, Pegg D, Faddy M, Gosden R. Permeation of human ovarian tissue with cryoprotective agents in preparation for cryopreservation. Human Reproduction. 1998;**13**(2):376-380

[62] Meryman H. Cryoprotective agents. Cryobiology. 1971;**8**(2):173-183

[63] Jörns A, Tiedge M, Lenzen S, Munday R. Effect of superoxide dismutase, catalase, chelating agents, and free radical scavengers on the toxicity of alloxan to isolated pancreatic islets in vitro. Free Radical Biology and Medicine. 1999;**26**(9):1300-1304

[64] Lakey JRT, Aspinwall CA, Cavanagh TJ, Kennedy RT. Secretion from islets and single islet cells following cryopreservation. Cell Transplantation. 1999;**8**(6):691-698

[65] Korbutt GS, Rayat GR, Ezekowitz J, Rajotte RV. CRYOPRESERVATION OF RAT PANCREATIC ISLETS: Effect of ethylene glycol on islet function and cellular Composition1. Transplantation. 1997;**64**(7):1065-1070

[66] Woods EJ, Liu J, Zieger MAJ, Lakey JRT, Critser JK. Water and cryoprotectant permeability characteristics of isolated human and canine pancreatic islets. Cell Transplantation. 1999;**8**(5):549-559

[67] Sasamoto H, Futami M, Ando Y, Nakaji S. Cryopreservation of rat islets of Langerhans by vitrification. Journal of Artificial Organs: The Official Journal of the Japanese Society for Artificial Organs. 2012;**15**(3):283-289

[68] Kojayan G, Whaley D, Alexander M, Rodriguez S, Lee S, Lakey JRT. Improved cryopreservation yield of pancreatic islets using combination of lower dose permeable cryoprotective agents. Cryobiology. 2019;**88**:23-28

[69] Beattie GM, Crowe JH, Lopez AD, Cirulli V, Ricordi C, Hayek A. Trehalose: A Cryoprotectant that enhances

recovery and preserves function of human pancreatic islets after Long-term storage. Diabetes. 1997;**46**(3):519-523

[70] Matsumoto S, Matsusita M, Morita T, et al. Effects of synthetic antifreeze glycoprotein analogue on islet cell survival and function during cryopreservation. Cryobiology. 2006;**52**(1):90-98

[71] Elliott GD, Wang S, Fuller BJ. Cryoprotectants: A review of the actions and applications of cryoprotective solutes that modulate cell recovery from ultra-low temperatures. Cryobiology. 2017;**76**:74-91

[72] El-Shewy HM, William FK, Darrabie M, Collins BH, Opara EC. Polyvinyl Pyrrolidone: A novel Cryoprotectant in islet cell cryopreservation. Cell Transplantation. 2004;**13**(3):237-243

[73] Baust JM. Molecular mechanisms of cellular demise associated with cryopreservation failure. Cell Preservation and Technology. 2002;**1**(1):17-31

[74] Bissoyi A, Nayak B, Pramanik K, Sarangi SK. Targeting cryopreservation-induced cell death: A review. Biopreservation and Biobanking. 2014;**12**(1):23-34

[75] Modak MA, Parab PB, Ghaskadbi SS. Pancreatic islets are very poor in rectifying oxidative DNA damage. Pancreas. 2009;**38**(1):23-29

[76] Robertson RP. Chronic oxidative stress as a central mechanism for glucose toxicity in pancreatic islet beta cells in diabetes. Journal of Biological Chemistry. 2004;**279**(41):42351-42354

[77] Janjic D, Andereggen E, Deng S, et al. Improved insulin secretion of cryopreserved human islets by antioxidant treatment. Pancreas. 1996;**13**(2):166-172

[78] Yamamoto I, Suga S, Mitoh Y, Tanaka M, Muto N. Antiscorbutic activity of L-ascorbic acid 2-glucoside and its availability as a vitamin C supplement in normal rats and Guinea pigs. Journal of Pharmacobio-Dynamics. 1990;**13**(11):688-695

[79] Arata T, Okitsu T, Fukazawa T, et al. Maintenance of glucose-sensitive insulin secretion of cryopreserved human islets with University of Wisconsin solution and ascorbic acid-2 glucoside. Artificial Organs. 2004;**28**(6):529-536

[80] Southard J, Ametani M, Vreugdenhil P, Lindell S, Pienaar B, Belzer F. Important components of the UW solution. Transplantation. 1990;**49**(2):251-257

[81] Hsuuw YD, Chang CK, Chan WH, Yu JS. Curcumin prevents methylglyoxal-induced oxidative stress and apoptosis in mouse embryonic stem cells and blastocysts. Journal of Cellular Physiology. 2005;**205**(3):379-386

[82] Huxtable R. Physiological actions of taurine. Physiological Reviews. 1992;**72**(1):101-163

[83] Soltani N, Qiu H, Aleksic M, et al. GABA exerts protective and regenerative effects on islet beta cells and reverses diabetes. Proceedings of the National Academy of Sciences. 2011;**108**(28):11692-11697

[84] Braun M, Ramracheya R, Bengtsson M, et al. γ-Aminobutyric acid (GABA) is an autocrine excitatory transmitter in human pancreatic β-cells. Diabetes. 2010;**59**(7):1694-1701

[85] Organization WH. The Selection and Use of Essential Medicines: Report of the WHO Expert Committee, 2015 (Including the 19th WHO Model List of Essential Medicines and the 5th WHO Model List of Essential Medicines for Children). Vol. 994.

Geneva, Switzerland: World Health Organization; 2015

[86] Hundal RS, Inzucchi SE. Metformin. Drugs. 2003;**63**(18):1879-1894

[87] Sirtori CR, Franceschini G, Galli-Kienle M, et al. Disposition of metformin (N, N-dimethylbiguanide) in man. Clinical Pharmacology & Therapeutics. 1978;**24**(6):683-693

[88] Chandravanshi B, Datar S, Bhonde R. Response of chick B islets to insulin secretagogues is comparable to those of human islet equivalents. JOP Journal of the Pancreas. 2015;**16**(3):318-323

[89] Morikawa M, Kimura T, Murakami M, Katayama K, Terada S, Yamaguchi A. Rat islet culture in serum-free medium containing silk protein sericin. Journal of Hepato-Biliary-Pancreatic Surgery. 2009;**16**(2):223

[90] Sasaki M, Kato Y, Yamada H, Terada S. Development of a novel serum-free freezing medium for mammalian cells using the silk protein sericin. Biotechnology and Applied Biochemistry. 2005;**42**(2):183-188

[91] de Vos P, Faas MM, Strand B, Calafiore R. Alginate-based microcapsules for immunoisolation of pancreatic islets. Biomaterials. 2006;**27**(32):5603-5617

[92] Fritschy WM, Wolters GH, Van Schilfgaarde R. Effect of alginate-polylysine-alginate microencapsulation on in vitro insulin release from rat pancreatic islets. Diabetes. 1991;**40**(1):37-43

[93] O'Shea GM, Sun AM. Encapsulation of rat islets of Langerhans prolongs xenograft survival in diabetic mice. Diabetes. 1986;**35**(8):943-946

[94] Chen W, Shu Z, Gao D, Shen AQ. Sensing and sensibility: Single-islet-based quality control assay of cryopreserved pancreatic islets with functionalized hydrogel microcapsules. Advanced Healthcare Materials. 2016;**5**(2):223-231

[95] Fraker C, Timmins MR, Guarino RD, et al. The use of the BD oxygen biosensor system to assess isolated human islets of langerhans: Oxygen consumption as a potential measure of islet potency. Cell Transplantation. 2006;**15**(8-9):745-758

[96] Kojayan GFA, Li S, Alexander M, Lakey JR. Cryopreserved alginate encapsulated islets can restore euglycemia in diabetic animal model better than cryopreserved non. Journal of Cell Medicine. 2019;**II**:1

[97] Malpique R, Osório LM, Ferreira DS, et al. Alginate encapsulation as a novel strategy for the cryopreservation of neurospheres. Tissue Engineering Part C: Methods. 2010;**16**(5):965-977

[98] Matsunari H, Maehara M, Nakano K, et al. Hollow fiber vitrification: A novel method for vitrifying multiple embryos in a single device. Journal of Reproduction and Development. 2012;**58**:2011-2051

[99] Maehara M, Matsunari H, Honda K, et al. Hollow fiber vitrification provides a novel method for cryopreserving in vitro maturation/fertilization-derived porcine embryos. Biology of Reproduction. 2012;**87**(6):131-138

[100] Nagaya M, Matsunari H, Kanai T, et al. An effective new cryopreservation procedure for pancreatic islets using hollow fiber vitrification. Hormone and Metabolic Research. 2016;**48**(08):540-549

[101] Giwa S, Lewis JK, Alvarez L, et al. The promise of organ and tissue preservation to transform medicine. Nature Biotechnology. 2017;**35**(6):530

Chapter 7

Cryopreservation in Ophthalmology

Yuting Shao, Chao Chen, Qi Zhou, Jun Yang, Xiao Lv, Mingyue Lin and Yanlong Bi

Abstract

Amniotic membranes (AMs) and corneas are critical materials in ocular surface reconstruction. AM has specific structures (e.g., basement and two types of cells with stemness characteristics: amniotic epithelial cells and amniotic mesenchymal cells), which contribute to its attractive physical and biological properties that make it fundamental to clinical application. The corneal endothelial cell is a vital part of the cornea, which can influence postoperative vision directly. However, widespread use of fresh AM and cornea has been limited due to their short use span and safety concerns. To overcome these concerns, different preservation methods have been introduced. Cryopreservation is distinguished from many preservation methods for its attractive advantages of prolonged use span, optimally retained tissue structure, and minimized infection risk. This review will focus on recent advances of cryopreserved AM and cornea, including different cryopreservation methods and their indications in ophthalmology.

Keywords: amnion, cornea, cryopreservation, indication, ophthalmology

1. Cryopreservation of amniotic membranes

1.1 Introduction

Ever since Davis [1] first used human amniotic membrane (**Figure 1**) (AM) for skin transplantation, people have been exploring this remarkable biomaterial. AM is located in the innermost layer of the fetal membranes [2]. It is 0.02–0.05 mm thick, lightweight, elastic, almost transparent, and avascular membrane classically composed of three layers: the epithelium, the basement membrane and the stroma layer [2]. There are two types of cells with stemness properties in AM: amniotic epithelial cells (AECs) and amniotic mesenchymal cells (AMSCs) [3], which are responsible for its unique biological properties including anti-inflammatory, anti-scarring, anti-microbial, angio-modulating, immunomodulatory, and anti-cancer effects [4–10]. Due to these properties, AM has become an ideal material for ocular reconstruction including the treatment of persistent epithelial defects and non-healing corneal ulcers, corneal perforations and descemetoceles, bullous keratopathy, as well as corneal disorders with associated limbal stem cell deficiency, pterygium, conjunctival reconstruction, corneoscleral melts and perforations, and glaucoma surgeries. However, its use span is short and many viruses (such as HIV-1/2, hepatitis B, hepatitis C, human T-cell lymphotropic virus, syphilis, and cytomegalovirus)

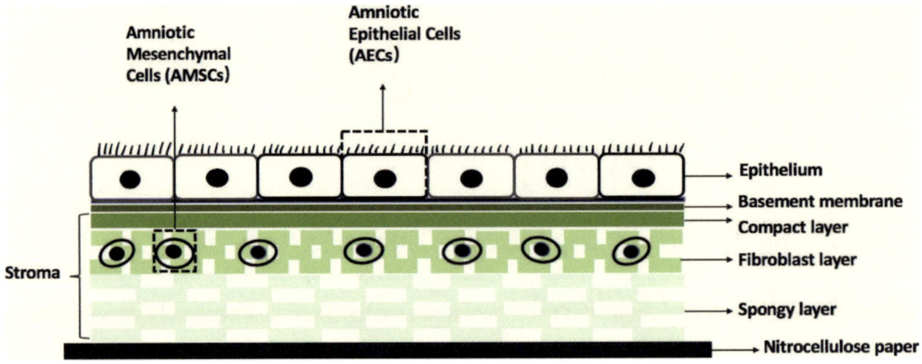

Figure 1.
Amniotic membrane (AM).

can be in their "window period" and escape detection, further limiting the use of fresh AM. To overcome these concerns, different preservation methods have emerged, such as freezing, lyophilization, and cryopreservation. However, most methods result in the destruction of endogenous cells and cause varying degrees of extracellular matrix (ECM) damage, which can affect the functionality of AM and its clinical benefits for wound treatment [11, 12]. Cryopreservation was first introduced by Lee and Tseng and has been proven to achieve high success rate in AM transplantation, which has been distinguished from many methods for its attractive advantages of prolonging use span, optimally retaining tissue structure, and minimizing infection risk [13, 14].

In this part, we classify the cryopreservation methods applied to amnion by commonly used cryoprotectant and analyze the influence of cryopreservation on AM combined with specific clinical trials.

1.2 General cryopreservation techniques

The AM is normally washed using balanced saline solution containing antibiotics such as streptomycin, penicillin, neomycin, and amphotericin prior to storage. Pieces of AM resting on a carrier are placed in a vial containing cryoprotectant solution at a controlled cooling rate. Storage temperatures of −80°C are often utilized, with the maximum storage times ranging between 1 and 2 years [1, 11, 12].

The main disadvantage of cryopreservation is the requirement of a deep-freezing facility, which is expensive, cumbersome, and frequently unavailable, especially in underdeveloped countries. In addition, maintaining stable storage temperatures during transportation is also relatively difficult.

1.3 Cryopreservation methods on AM

1.3.1 Glycerol-cryopreservation

Glycerol storage was first introduced in the Netherlands in 1984 to preserve donor skin for transplantation [13]. Positive results over subsequent decades have led to its clinical acceptance, including in the preservation of AM. Glycerol has led to higher cell viability and higher bFGF secretion for up to three months of AM storage [14]. After strict preservation and sterilization processes, pieces of AM resting on a carrier are placed in a vial containing storage solution. Tseng's laboratory introduced a methodology of glycerol (86%) in Dulbecco's Modified Eagle Medium at a ratio of 1:1 [15, 16]. The most common cryopreservation protocol reported in the literature involves the use of 50% glycerol and storage at −80°C [17–21]. Undiluted and 98% glycerol have also been reported to be clinically effective [15].

In 2011, Thomasen et al. [21] showed that long-term storage of 50% glycerol cryopreserved AM for durations up to 24 months at −80°C did not significantly impair the histology of AM. Wagner et al. [14] used 85% glycerol for cryopreserved AM, and their histological examinations had no significant alterations following cryopreservation, either for straight cryopreservation or with glycerol. They also demonstrated that neither tensile strength nor Young's modulus was significantly influenced by the storage method. In addition, they also detected a significant increase in tensile strength over storage time, independent of the storage method.

Some groups have found that storage of AM in 50% glycerol at −80°C decellularizes the AM and results in low viability [17–20]. Interestingly, the results from Wagner et al. [14] research showed that epithelial cells were not significantly reduced during freezing in comparison to stromal cells, possibly indicating a higher sensitivity of stromal AM cells to freezing damage than epithelial cells (**Figure 2**). Through repeated measurement analysis, storage time showed a significant effect on cell viability. Prabhasawat et al. [22, 23] reported that the use of a highly concentrated glycerol solution abolishes AM cell viability. The possible toxic effect of glycerol is responsible for that.

To summarize, glycerol-cryopreserved AM retains the histological characteristics of fresh AM independent of an increase in glycerol concentration. Tensile strength and elasticity can also be better preserved, with tensile strength increasing with storage time. However, the cell viability of cryopreserved AM was significantly affected by storage time and glycerol concentration. In particular, the stromal cells were more sensitive. A previous study [24] showed that this method had little effect on the growth factors of AM. More research is needed to confirm the effect of glycerol cryopreservation on AMs.

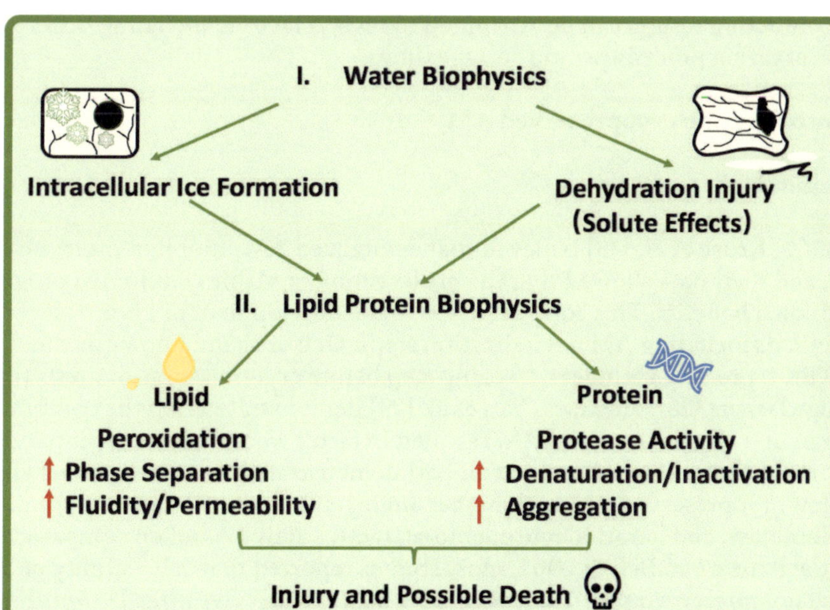

Figure 2.
Pathways of cellular injury during freeing.

1.3.2 DMSO-cryopreservation

DMSO has been used as an alternative for AM in glycerol-cryopreservation. An increasing concentration of DMSO is used instead of washing the AM with an antibiotic-saline solution after placenta collection [12]. Azuara-Blanco et al. [25] used 4%, 8%, and 10% DMSO, while Kubo et al. [26] used 0.5 M, 0.1 M, and 0.15 M DMSO for washing. AMs can be stored in 10% or 0.15 M DMSO at −80°C for several months without significant damage. In general, solutions containing DMSO are used less often for AM cryopreservation compared to glycerol, due to high toxicity [12]. However, AM storage solutions containing DMSO have been studied a lot regarding its ability to increase cell viability in AM under experimental conditions [2].

A cryopreservation method with DMSO from Duan-Arnold's group [24] showed a retained cell viability of over 80%. Cryopreserved AM tested after three months of storage showed no changes in the tissue architecture and collagen IV, which exists in the basement membrane, compared with fresh AM. However, in 2015, Yazdanpanah et al. [8] showed that the viability of epithelial cells in fresh AMs was estimated at 97% after staining with trypan blue, decreasing to about 50% in DMSO cryopreserved tissues after six months. They evaluated the effects of cryopreservation on AM angiogenesis modulation activity compared to fresh tissue in an animal model, showing that cryopreserved AM has the same effect on angiogenesis as fresh AM. The epithelial surface of cryopreserved AM inhibited angiogenesis, and the mesenchymal surface augmented vessel sprouting and length. In 2013, Tehrani et al. [27] used 10% DMSO as a cryoprotectant to evaluate the antibacterial properties of AM after preservation in vitro. The results of this study showed that the antibacterial property of AM was maintained after cryopreservation, but was dependent on bacterial genus and strain.

To sum up, the literature we collected on DMSO-cryopreserved AM showed no significant differences in tissue integrity and biological properties (antibacterial and angiogenesis modulation) compared with fresh AM. However, although many research groups use DMSO as a cryoprotectant, the data related to cell viability vary. These conflicting results can be attributed to several factors, including differing cryopreservation procedures and storage times.

1.4 Controversy on cryopreserved AM

1.4.1 Variable cell viabilities

In 2000, Kruse et al. [18] believed that devitalized AM exhibited therapeutic effects, and their data showed that the preservation of viable cells in AM provided no additional benefits. This conclusion led to the development of cryopreservation methods including AM devitalization steps. One of them, known as the CRYOTEK® process, includes a freezing step before cryopreservation, resulting in devitalized tissue [28]. However, Yan et al. [29] demonstrated that the combination of exogenous cells and acellular AM resulted in faster wound closure compared with acellular AM alone. Duan-Arnold et al. [24] demonstrated that endogenous viable cells allow cryopreserved AM with higher angiogenic, anti-inflammatory, antioxidant, fibroblast, and keratinocyte chemo-attractive activities when compared with AM in devitalization. Before 2001, most studies reported that cell viability of 50% or less at cryopreserved post-thaw with cells failing to survive after 18 months of storage at −80°C [26, 30]. Since then, scientists have been attempting to improve the cryopreservation method, for improved cell viability retention. For example, the cryopreservation protocol invented by the group of Duan Arnold et al. [24]

can maintain 70% or greater cell viability after 24 months of storage at −80°C. AM storage solutions containing dimethyl sulfoxide (DMSO) have been studied, mostly under experimental conditions, and shown the ability to increase AM cell viability [2]. Although the survival of amniotic cells is related to storage time, different cryopreservation steps can also affect cell viability, thus exerting different clinical effects.

1.4.2 Storage temperature

The best storage temperature (−196°C or − 80°C) is also a controversial issue for cryopreservation. AMs stored at −196°C have showed morphology similar to fresh AM in both preservation media, and AM stored at −80°C showed disruption of the stromal matrix [2, 31]. However, −80°C is still widely used by international scientists.

To sum up, cryopreservation protocols are not standardized. Preparation and sterilization before cryopreservation, as well as the selection of cryoprotectant during cryopreservation, will lead to high variability in cell viability [2, 32]. Different storage temperature and storage time also affects the structure and function of amniotic membrane. It is important to establish adaptable protocols for the clinical banking of AM that include verification of graft quality and viability before its release for transplantation, whether in the trial or clinical stage.

1.5 Commercially available cryopreserved AM

1.5.1 PROKERA®Slim (PKS)/PROKERA®(PK)

PROKERA®Slim (PKS) (Bio-Tissue, Inc., Miami, FL, USA) is a Class II medical device approved by the Food and Drug Administration in 2003 to be used as a temporary AM patch for delivering the biological actions of AM to a corneal surface without using sutures. It contains a piece of cryopreserved AM clipped into a concave poly-carbonate dual-ring system, like a symblepharon ring, that conforms to the corneal and limbal surface like a contact lens. The ring system has an inner diameter of 15 or 16 mm.

It has become the most common commercially available cryopreserved AM product in ophthalmology and is applied to various ocular surface and orbital disorders. It is a safe and effective method that makes AM transplantation sutureless and adhesiveless, contributing to healing and reconstruction of the ocular surface and orbit with minimal side effects [33]. However, PROKERA is not recommended for eyes with functioning blebs or glaucoma drainage implants because of the oppositional positioning of the retaining ring [34].

2. Cryopreservation of cornea

2.1 Introduction

Corneal disease is one of the world's leading causes of blindness. Corneal scarring and haze due to various factors can affect vision, making corneal transplantation an important means of treatment for corneal diseases [35–37]. Advances in corneal preservation techniques have improved the survival rate of corneal grafts [38] and have largely contributed to the development of modern corneal transplant surgery [39]. With the flourishing of corneal preservation technology, breakthroughs have been made in preservation times and corneal activity. Nevertheless, cryopreservation is the only current method that can virtually preserve tissue structure for a long time.

Meanwhile, the development of modern eye banks have been accompanied by the advancement of corneal preservation technology. The establishment of an eye bank provides favorable conditions for corneal transplantation [40, 41].

2.2 Corneal transplantation and preservation

The idea of replacing the turbid cornea with transparent tissue was first proposed by Pellier de Quengsy in 1789 [42, 43]. In 1824, Reisinger exploited animal corneas in surgery [44], which was named keratoplasty. Later in the nineteenth century, a large number of animal experiments helped doctors realize that inter-species transplantation was a necessary condition to avoid corneal opacity after transplantation [45–47]. In view of this, researchers began to experiment with human corneal transplantation. Early corneal transplantation relied on living donor tissues due to fears relating to transplanting dead tissue. The first successful full-thickness corneal transplantation (including all corneal layers) was completed in 1905 [48]. It was not until the 1930s that the cornea of the deceased donor was used and the entire eye was kept in a glass jug (wet room) for several days [49].

In 1912, Magitot reported that excised corneal grafts could be preserved in red blood cells at 5°C for eight days [50–52]. The grafts were successfully used for corneal lamellar transplantation [53]. At first, the freshness of the cornea was considered key to corneal transplantation [54, 55]. However, Ukrainian doctor Filatov systematically reported the application of corpus corneal tissue to clinical practice [56], which possessed an inter-generational meaning. It opened a new era of corneal preservation and transplantation [57–59]. These developments led to the establishment of the world's first ophthalmology bank in New York in the 1940s. The preservation technique of the original eye bank was very simple [60, 61]: eyeballs were kept in a small glass bottle in a humid and cool environment [62]. Immediate removal of the eyeball after donor death was the only way to ensure the quality of the corneal grafts [63–65]. Eye banks were established in major cities, such as London, to guarantee that eyeballs were promptly forwarded [66, 67]. In the early 1950s, the activity of CECs was first considered as an important factor affecting transplantation [68–70]. The emphasis on preservation techniques was transferred to maintain the activity and integrity of CECs [71, 72]. Since then, corneal preservation techniques have been increasingly successful, resulting in approximately 40,000 corneal transplants per year in the United States, 20,000 per year in Europe, and thousands per year in other countries, such as India.

2.3 Corneal preservation methods

Corneal preservation is divided into two categories according to the survival of CECs: inactive and active preservation [73–75]. The former method includes dry preservation and cryopreservation [76–78] and operates under the principle of removing corneal moisture while inhibiting enzyme activity and autolysis in cells for long-term preservation [79, 80]. Common preservatives are glycerin, molecular sieves, and silica gel [81–83], which can preserve intact lamellar collagen structure [84]. Active preservation comprises short-term (hours to two days), medium-long term (7 to 30 days) and long-term (months to years) preservation. In terms of storage conditions, it utilizes normal (34~37°C), low (usually 4°C) and deep low (subzero) temperatures [85–88].

Short-term corneal storage mainly refers to the preservation of wet rooms, the simplest and most convenient of all corneal storage technologies. For this reason, it is still the basic technology for preserving cornea in the eye banks of developing countries. As for medium-term corneal preservation, corneal preservation solution is stored at 4°C for 4 to 14 days [89].

The prolongation of corneal preservation allows more preparation for patients and flexible adjusting of operation times, while also satisfying blood test and corneal transportation times. With the improvement of preservation techniques, the composition of the corneal preservation solution has been constantly changing. A certain concentration of chondroitin sulfate is added to modify M-K solution, which can alleviate the swelling state during preservation. Optisol corneal medium preservation solution was proposed by Lindstrom and has become the most common preservation solution in US eye banks, which is mainly a mixture of K-liquid and Dexsol solution [90]. Long-term corneal preservation refers to organ culture storage and cryopreservation. Organ culture is to simulate the presence of a normal human cornea environment with medium at 30–37°C [91].

At present, there are several corneal preservation methods applied in global eye banks, but none of those is perfect. Each preservation method has its own advantages and disadvantages, which differ from the preservation temperature, the composition of the preservation solution, and the penetrant preventing matrix edema.

2.4 Cryopreservation

After donor death, the sudden stop of the aqueous humor causes nutrient and oxygen shortages, leading to final depletion at room temperature, which can, in turn, lead to autolysis of the corneal cells and initial damage to the cornea [92]. During the period from donor death to corneal removal and storage, the donor's corpse is exposed to room temperatures, necessitating minimal time delays to ensure that the initial donor cornea is healthy and intact along with functional endothelial cells.

The acceptable short storage time, as well as organ damage, poses a logistical challenge to organ storage and ultimately affects grafts and patient survival. Prolonged storage times can cause many transplantable organs, further exacerbating the growing imbalance between organ supply and demand. Organ cryopreservation is used to preserve long-lived cells and tissues. Theoretically, the storage of biological materials, including cells, tissues, and organs for transplantation at a low temperature (i.e., in liquid nitrogen at −160°C) is uncertain [93, 94]. Such a technique would have the potential to alter the way in which organs are recovered, distributed, and utilized for transplantation. However, ice is the biggest enemy in the cryopreservation of organs and tissues. Ice crystals, especially intracellular ice, can cause significant cellular damage and destroy the complex macroscopic tissues of intact organs. In this field, current developments are used to avoid the formation of ice, or mitigate it, during cryogenic storage. Any successful organ cryopreservation strategy requires a delicate balance between the relative needs of cryopreservation and toxicity in these situations.

2.5 Corneal cryopreservation technology

In 1954, Eastcott first adopted a cryopreserved human cornea for transplantation successfully [95, 96], pretreating the keratin tissue in glycerol before freezing it in a mixture of alcohol and carbon dioxide for cryopreservation of the full-thickness cornea [97]. This method generally removes the cornea under the protection of a cryogen to −80°C, and stores it in liquid nitrogen at −196°C. Therefore, the CECs are in a dormant state. The state can completely inhibit the metabolism of cells, eliminate the toxic effects caused by the accumulation of metabolites, and avoid the need to change the liquid during organ culture. In addition, it also restrains microorganisms during cryopreservation, protecting the cornea from microbial invasion.

The components currently contained in corneal cryoprotectants include DMSO, propylene glycol, ChS, and sucrose. DMSO is a relatively stable protective agent to maintain the integrity of corneal cells, while sucrose molecules act as buffers in corneal protection, and ChS improves CEC activity in cryopreservation [98]. DMSO began to be treated as a tissue preservative to preserve cultured rabbit CECs by Smith [99]. Shortly thereafter, Mueller injected a preservation solution containing DMSO into the anterior chamber of an eyeball, placing the eyeball in a preservation solution containing glycerol. The cornea was removed before surgery for full-thickness transplantation [100, 101]. In 1965, Capella [102] used DMSO as an antifreeze to improve a cryopreservation method, ensuring corneal graft activity. According to another report [103], the clinical application of cryopreservation techniques has little differences in techniques. The corneal tissue must be preserved eight hours after death. By increasing the level of DMSO, it eventually reaches a concentration of 7.5%. The classic four-step cooling is to initiate a cooling rate at 1.5~2°C/min, drop the temperature to −30°C, change to 5–7°C/min, and ultimately maintain −80°C [104, 105].

It is still essential to further explore the rate of cooling to keep CEC activity and reduce cell loss [106, 107]. Temperature-controlled thawing before transplantation is a key step in protecting the corneal endothelium. At present, the prevalent view is that rapid rewarming could decrease the contact of cells with high concentrations of electrolytes and reduce cell damage [108]. The thawing process of the cryopreserved cornea must be strictly controlled, as the solute containing DMSO has endotoxicity once the temperature exceeds 37°C [109]. Cryopreservation would impair the morphology and function of the corneal endothelium. During the thawing process, an ascending solute concentration, the formation of crystals, changes in pH, and osmotic pressure will reduce the survival rate of CECs [110]. Glycerol, polyvinylpyrrolidone, and DMSO can all be used as cryopreservatives, but DMSO is currently the most widely used [111, 112].

2.6 Effect of corneal cryopreservation

The ultra-low temperature preservation method overcomes the drawbacks of most other corneal preservation methods, significantly prolonging corneal preservation time, reducing pollution, and avoiding the toxic effects of its own metabolic substances. Electron microscopy can observe changes in the subcellular morphology of CECs caused by cryopreservation, some of which are considered irreversible [113]. Studies have shown that, after cryopreservation, the barrier function of endothelial cells is impaired. Compared with wet room preservation and MK solution preservation, cryopreserved corneal grafts have been completely transparent for a long time after surgery. For one-year cryopreservation, 55% of endothelial cells were deactivated, while the rate of CECs preserved by MK solution was only 21–22% [114]. There are barely significant structural differences in microbiological, histological, and ultrastructural features when comparing long-term cryopreservation of tissue (>7 years) and short-term cryopreserved cat corneal sclera (<1 year) [115]. As such, tissues cryopreserved for up to 10 years could be used for tectonic support without structural or microbial barriers.

Under suitable conditions, no crystal solidification occurs during the freezing process, called vitrification [116]. Vitrification requires a high concentration of cryoprotectant, yet theoretically, tissue could be stored in a very low temperature environment without forming intracellular or extracellular crystals, and corneal endothelium damage could be avoided significantly [117]. Glycerol, 1,2-propanetriol, and 2,3-butanediol are all considered as eligible cryopreservation agents for corneal vitrification [118, 119].

Studies have found that an effective concentration of a single cryopreservative is toxic to CECs, yet the mixture of preservatives or the addition of preservatives at low temperatures seems to reduce toxicity [120]. As a means of corneal preservation, further study is warranted to investigate whether vitrification would achieve good results. In 1981, Sperling used corneal grafts in a corneal preservation solution at the early stage and carried out a cryopreservation operation later. After rewarming, the cornea was transferred to a preservation solution, identifying corneal activity. The following studies indicated that the corneal grafts maintained transparency 71% of the time after 1 year and 58% of the time after 12 years [121].

In our previous study, we performed lamellar keratoplasty combined with keratopigmentation in 22 corneal leukoma eyes using glycerol-cryopreserved corneal tissues, and no graft-rejection occurred during the 3 years of follow-up. Moreover, the outcome of a low graft rejection rate in glycerol-cryopreserved corneal tissues was also confirmed by our preceding study in treating Terrien marginal degeneration. In the subsequent study, for patients with refractive herpes simplex keratitis, 3 eyes of 27 eyes (11.1%) suffered allograft stromal rejection, all eyes reversed after prompt medication. Meanwhile, only 2 eyes (7.41%) exhibited refractive herpes simplex keratitis recurrence and the main site was located at the margin of the graft and the recipient bed. This result is consistent with the theory that grafts survive better when compared with reports clarifying that up to 33% of patients have suffered recurrence using fresh grafts. The recurrence rate in fresh grafts may be partially related to the long-term usage of topical steroid eye drops; however, it may be much more closely correlated with fewer keratocytes in the cryopreserved donor tissue to reactivate immune-inflammatory responses [122–124]. Based on the above information, glycerol-cryopreserved corneal tissues can be effectively and biosafely used with a low rejection and recurrence rate in corneal transplantation, especially in developing countries where good donor corneas are difficult to get.

3. Conclusion

The cryopreservation method can preserve the activity of the AM and cornea for extended periods up to several years, solving the problem of preservation time and activity deterioration. However, equipment complications, expensive technical support, and transport difficulties have become impediments to widespread use. The functional status of AM, endothelial cells, and corneal transparency have been of vital importance in the development of cryopreservation. As researchers become more aware of the function and properties of CECs, attempts to find a more conducive method and media for the preservation of AMs and corneas will continue.

Conflict of interest

We declare that we have no conflicts of interest.

Funding

1. The National Nature Science Foundation of China (81470028).

2. The Municipal Human Resources Development Program for Outstanding Leaders in Medical Disciplines in Shanghai (2017BR060).

3. Shanghai Shenkang Hospital Development Project: A Three- Year Action Plan (16CR3027A).

Author details

Yuting Shao†, Chao Chen†, Qi Zhou, Jun Yang, Xiao Lv, Mingyue Lin and Yanlong Bi*
Department of Ophthalmology, Tongji Hospital, Tongji University School of Medicine, Shanghai, China

*Address all correspondence to: biyanlong@tongji.edu.cn

† Co-first author.

References

[1] Dua HS, Gomes JA, King AJ, Maharajan VS. The amniotic membrane in ophthalmology. Survey of Ophthalmology. 2004;**49**(1):51-77. DOI: 10.1016/j.survophthal.2003.10.004

[2] Hettiarachchi D, Dissanayake VHW, Goonasekera HWW. Optimizing amniotic membrane tissue banking protocols for ophthalmic use. Cell and Tissue Banking. 2016;**17**(3):387-397. DOI: 10.1007/s10561-016-9568-3

[3] Koike C, Zhou KX, Takeda Y, Fathy M, Okabe M, Yoshida T, et al. Characterization of amniotic stem cells. Cellular Reprogramming. 2014;**16**(4):298-305. DOI: 10.1089/cell.2013.0090

[4] Shimmura S, Shimazaki J, Ohashi Y, Tsubota K. Antiinflammatory effects of amniotic membrane transplantation in ocular surface disorders. Cornea. 2001;**20**(4):408-413. DOI: 10.1097/00003226-200105000-00015

[5] Sorensen OE, Cowland JB, Theilgaard-Monch K, Liu LD, Ganz T, Borregaard N. Wound healing and expression of antimicrobial peptides/polypeptides in human keratinocytes, a consequence of common growth factors. Journal of Immunology. 2003;**170**(11):5583-5589. DOI: 10.4049/jimmunol.170.11.5583

[6] Niknejad H, Paeini-Vayghan G, Tehrani FA, Khayat-Khoei M, Peirovi H. Side dependent effects of the human amnion on angiogenesis. Placenta. 2013;**34**(4):340-345. DOI: 10.1016/j.placenta.2013.02.001

[7] Hao YX, Ma DHK, Hwang DG, Kim WS, Zhang F. Identification of antiangiogenic and antiinflammatory proteins in human amniotic membrane. Cornea. 2000;**19**(3):348-352. DOI: 10.1097/00003226-200005000-00018

[8] Yazdanpanah G, Paeini-Vayghan G, Asadi S, Niknejad H. The effects of cryopreservation on angiogenesis modulation activity of human amniotic membrane. Cryobiology. 2015;**71**(3):413-418. DOI: 10.1016/j.cryobiol.2015.09.008

[9] Talmi YP, Sigler L, Inge E, Finkelstein Y, Zohar Y. Antibacterial properties of human amniotic membranes. Placenta. 1991;**12**(3):285-288. DOI: 10.1016/0143-4004(91)90010-D

[10] Seo JH, Kim YH, Kim JS. Properties of the amniotic membrane may be applicable in cancer therapy. Medical Hypotheses. 2008;**70**(4):812-814. DOI: 10.1016/j.mehy.2007.08.008

[11] Jirsova K, Jones GLA. Amniotic membrane in ophthalmology: Properties, preparation, storage and indications for grafting—A review. Cell and Tissue Banking. 2017;**18**(2):193-204. DOI: 10.1007/s10561-017-9618-5

[12] Riau AK, Beuerman RW, Lim LS, Mehta JS. Preservation, sterilization and de-epithelialization of human amniotic membrane for use in ocular surface reconstruction. Biomaterials. 2010;**31**(2):216-225. DOI: 10.1016/j.biomaterials.2009.09.034

[13] Hermans MHE. Clinical-experience with glycerol-preserved donor skin treatment in partial thickness burns. Burns. 1989;**15**(1):57-59. DOI: 10.1016/0305-4179(89)90074-0

[14] Wagner M, Walter P, Salla S, Johnen S, Plange N, Rutten S, et al. Cryopreservation of amniotic membrane with and without glycerol additive. Graefes Archive for Clinical and Experimental Ophthalmology. 2018;**256**(6):1117-1126. DOI: 10.1007/s00417-018-3973-1

[15] Lee SH, Tseng SCG. Amniotic membrane transplantation for persistent epithelial defects with ulceration. American Journal of Ophthalmology. 1997;**123**(3):303-312. DOI: 10.1016/S0002-9394(14)70125-4

[16] Malhotra C, Jain AK. Human amniotic membrane transplantation: Different modalities of its use in ophthalmology. World Journal of Transplantation. 2014;**4**(2):111-121. DOI: 10.5500/wjt.v4.i2.111

[17] Cooke M, Tan EK, Mandrycky C, He H, O'Connell J, Tseng SCG. Comparison of cryopreserved amniotic membrane and umbilical cord tissue with dehydrated amniotic membrane/chorion tissue. Journal of Wound Care. 2014;**23**(10):465. DOI: 10.12968/jowc.2014.23.10.465

[18] Kruse FE, Joussen AM, Rohrschneider K, You LT, Sinn B, Baumann J, et al. Cryopreserved human amniotic membrane for ocular surface reconstruction. Graefe's Archive for Clinical and Experimental Ophthalmology. 2000;**238**(1):68-75. DOI: 10.1007/s004170050012

[19] Shortt AJ, Secker GA, Lomas RJ, Wilshaw SP, Kearney JN, Tuft SJ, et al. The effect of amniotic membrane preparation method on its ability to serve as a substrate for the ex-vivo expansion of limbal epithelial cells. Biomaterials. 2009;**30**(6):1056-1065. DOI: 10.1016/j.biomaterials.2008.10.048

[20] Niknejad H, Deihim T, Solati-Hashjin M, Peirovi H. The effects of preservation procedures on amniotic membrane's ability to serve as a substrate for cultivation of endothelial cells. Cryobiology. 2011;**63**(3):145-151. DOI: 10.1016/j.cryobiol.2011.08.003

[21] Thomasen H, Pauklin M, Noelle B, Geerling G, Vetter J, Steven P, et al. The effect of long-term storage on the biological and histological properties of cryopreserved amniotic membrane. Current Eye Research. 2011;**36**(3):247-255. DOI: 10.3109/02713683.2010.542267

[22] Prabhasawat P, Kosrirukvongs P, Booranapong W, Vajaradul Y. Application of preserved human amniotic membrane for corneal surface reconstruction. Cell and Tissue Banking. 2000;**1**(3):213-222. DOI: 10.1023/A:1026542702099

[23] Prabhasawat P, Tseng SCG. Impression cytology study of epithelial phenotype of ocular surface reconstructed by preserved human amniotic membrane. Archives of Ophthalmology-Chic. 1997;**115**(11):1360-1367. DOI: 10.1001/archopht.1997.01100160530001

[24] Duan-Arnold Y, Gyurdieva A, Johnson A, Uveges TE, Jacobstein DA, Danilkovitch A. Retention of endogenous viable cells enhances the anti-inflammatory activity of cryopreserved amnion. Advances in Wound Care. 2015;**4**(9):523-533. DOI: 10.1089/wound.2015.0636

[25] Azuara-Blanco A, Pillai CT, Dua HS. Amniotic membrane transplantation for ocular surface reconstruction. British Journal of Ophthalmology. 1999;**83**(4):399-402. DOI: 10.1136/bjo.83.4.399

[26] Kubo M, Sonoda Y, Muramatsu R, Usui M. Immunogenicity of human amniotic membrane in experimental xenotransplantation. Investigative Ophthalmology & Visual Science. 2001;**42**(7):1539-1546

[27] Tehrani FA, Ahmadiani A, Niknejad H. The effects of preservation procedures on antibacterial property of amniotic membrane. Cryobiology. 2013;**67**(3):293-298. DOI: 10.1016/j.cryobiol.2013.08.010

[28] Amniox Medical, Inc. CRYOTEK process. Available from: www.amnioxmedical.com/CRYOTEK-process.html

[29] Yan GH, Sun HQ, Wang F, Wang JP, Wang FC, Zou ZM, et al. Topical application of hPDGF-A-modified porcine BMSC and keratinocytes loaded on acellular HAM promotes the healing of combined radiation-wound skin injury in minipigs. International Journal of Radiation Biology. 2011;**87**(6):591-600. DOI: 10.3109/09553002.2011.570854

[30] Rama P, Giannini R, Bruni A, Gatto C, Tiso R, Ponzin D. Further evaluation of amniotic membrane banking for transplantation in ocular surface diseases. Cell and Tissue Banking. 2001;**2**(3):155-163. DOI: 10.1023/A:1020158206073

[31] Tan EK, Cooke M, Mandrycky C, Mahabole M, He H, O'Connell J, et al. Structural and biological comparison of cryopreserved and fresh amniotic membrane tissues. Journal of Biomaterials and Tissue Engineering. 2014;**4**(5):379-388. DOI: 10.1166/jbt.2014.1180

[32] Paolin A, Trojan D, Leonardi A, Mellone S, Volpe A, Orlandi A, et al. Cytokine expression and ultrastructural alterations in fresh-frozen, freeze-dried and gamma-irradiated human amniotic membranes. Cell and Tissue Banking. 2016;**17**(3):399-406. DOI: 10.1007/s10561-016-9553-x

[33] Pachigolla G, Prasher P, Di Pascuale MA, McCulley JP, McHenry JG, Mootha VV. Evaluation of the role of ProKera in the management of ocular surface and orbital disorders. Eye & Contact Lens. 2009;**35**(4):172-175. DOI: 10.1097/ICL.0b013e3181a66a12

[34] Shupe AT, Cheng AMS. Case report: Use of Self-retained cryopreserved amniotic membrane in the treatment of acute chemical ocular burn. Optometry and Vision Science. 2017;**94**(11):1062-1065. DOI: 10.1097/OPX.0000000000001132

[35] Ayatollahi H, Nourani A, Khodaveisi T, Aghaei H, Mohammadpour M. Teleophthalmology in practice: Lessons learned from a pilot project. The Open Medical Informatics Journal. 2017;**11**:20-28. DOI: 10.2174/1874431101711010020

[36] Tseng SS, Yang WW, Hsiao JC. Bilateral corneal erosion due to retail purchase of unfitted prescription contact lenses: A case report. Cornea. 2008;**27**(10):1179-1181. DOI: 10.1097/ICO.0b013e318173140e

[37] Suh LH, Zhang C, Chuck RS, Stark WJ, Naylor S, Binley K, et al. Cryopreservation and lentiviral-mediated genetic modification of human primary cultured corneal endothelial cells. Investigative Ophthalmology & Visual Science. 2007;**48**(7):3056-3061. DOI: 10.1167/iovs.06-0771

[38] Chang KH, Tan HP, Kuo CC, Kuo DH, Shieh P, Chen FA, et al. Effect of nortriptyline on Ca^{2+} handling in SIRC rabbit corneal epithelial cells. The Chinese Journal of Physiology. 2010;**53**(3):178-184

[39] Yeh JH, Sun TK, Chou CT, Chen WC, Lee JK, Yeh HC, et al. Effect of sertraline on Ca^{2+} fluxes in rabbit corneal epithelial cells. The Chinese Journal of Physiology. 2015;**58**(2):85-94. DOI: 10.4077/CJP.2015.BAC255

[40] Horner IJ, Kraut ND, Hurst JJ, Rook AM, Collado CM, Atilla-Gokcumen GE, et al. Effects of polyhexamethylene biguanide and polyquaternium-1 on phospholipid bilayer structure and dynamics. The Journal of Physical Chemistry. B. 2015;**119**(33):10531-10542. DOI: 10.1021/acs.jpcb.5b07162

[41] Smith VA, Johnson T. Evaluation of an animal product-free variant of MegaCell MEM as a storage medium for corneas destined for transplantation.

Ophthalmic Research. 2010;**43**(1):33-42. DOI: 10.1159/000246576

[42] Pels E, Beele H, Claerhout I. Eye bank issues: II. Preservation techniques: warm versus cold storage. International Ophthalmology. 2008;**28**(3):155-163. DOI: 10.1007/s10792-007-9086-1

[43] Balarabe AH, Hassan R, Fatai OO. Eye health seeking habits and barriers to accessing curative services among blind beggars in an urban community in Northern Nigeria. Annals of African Medicine. 2014;**13**(4):184-188. DOI: 10.4103/1596-3519.142289

[44] Sanz-Marco E, Lopez-Prats MJ, Garcia-Delpech S, Udaondo P, Diaz-Llopis M. Fulminant bilateral *Haemophilus influenzae* keratitis in a patient with hypovitaminosis A treated with contaminated autologous serum. Clinical Ophthalmology. 2011;5:71-73. DOI: 10.2147/OPTH.S15847

[45] Saltzmann RM, Yep JM, Blomquist PH. Fusarium keratitis associated with ReNu with MoistureLoc sample kits. Eye & Contact Lens. 2008;**34**(6):337-339. DOI: 10.1097/ICL.0b013e31818c25bf

[46] Wilde C, Messina M, Moshiri T, Snape SE, Maharajan S. Interface *Scopulariopsis gracilis* fungal keratitis following Descemet's stripping automated endothelial keratoplasty (DSAEK) with a contaminated graft. International Ophthalmology. 2018;**38**(5):2211-2217. DOI: 10.1007/s10792-017-0706-0

[47] Montes JA, Johnson D, Jorgensen J, McElmeel ML, Fulcher LC, Kiel JW. Potency and sterility of fortified tobramycin, fortified vancomycin, and moxifloxacin at 4, 24, and 35°C for 14 days. Cornea. 2016;**35**(1):122-126. DOI: 10.1097/ICO.0000000000000676

[48] Armitage WJ, Tullo AB, Larkin DF. The first successful full-thickness corneal transplant: A commentary on Eduard Zirm's landmark paper of 1906. The British Journal of Ophthalmology. 2006;**90**(10):1222-1223. DOI: 10.1136/bjo.2006.101527

[49] Filatov VP. Perspective of optical transplantation of the cornea. Vestnik Oftalmologii. 1946;**25**(1):34-38

[50] John S, Sengupta S, Reddy SJ, Prabhu P, Kirubanandan K, Badrinath SS. The Sankara Nethralaya mobile teleophthalmology model for comprehensive eye care delivery in rural India. Telemedicine Journal and E-Health. 2012;**18**(5):382-387. DOI: 10.1089/tmj.2011.0190

[51] Fairbairn NG, Randolph MA, Redmond RW. The clinical applications of human amnion in plastic surgery. Journal of Plastic, Reconstructive & Aesthetic Surgery. 2014;**67**(5):662-675. DOI: 10.1016/j.bjps.2014.01.031

[52] Zimmerman AB, Richdale K, Mitchell GL, Kinoshita BT, Lam DY, Wagner H, et al. Water exposure is a common risk behavior among soft and gas-permeable contact lens wearers. Cornea. 2017;**36**(8):995-1001. DOI: 10.1097/ico.0000000000001204

[53] Rambaud C, Tabary A, Contraires G, El Hassan F, Labalette P. Contextual study of Airsoft gun related ocular injuries. Journal Français d'Ophtalmologie. 2013;**36**(3):236-241. DOI: 10.1016/j.jfo.2012.01.012

[54] Meizner I. Three dimensional (3D) ultrasound in gynecology: A breakthrough? Harefuah. 2007;**146**(4):274-275, 317-8

[55] John T. Human amniotic membrane transplantation: Past, present, and future. Ophthalmology Clinics of North America. 2003;**16**(1):43-65 vi

[56] Greenwald R. The handling of corneal donor tissue before penetrating

keratoplasty. Today's OR Nurse. 1994;**16**(5):33-35

[57] Adeghate E, Donath T. Transplantation of tissue grafts into the anterior eye chamber: A method to study intrinsic neurons. Brain Research Brain Research Protocols. 2000;**6**(1-2):33-39

[58] Firdova M, Tkacova E, Jesenska Z, Minova M. Mycotic keratitis with uveitis cause by *Scedosporium apiospermum*. Ceska a Slovenska Oftalmologie: Casopis Ceske Oftalmologicke Spolecnosti a Slovenske Oftalmologicke Spolecnosti. 1997;**53**(4):248-251

[59] Wang B, Wang Y, Ding J. Repair of cornea-sclera defect by autogenous sclera graft from the same eye. Zhongguo xiu fu chong jian wai ke za zhi = Zhongguo xiufu chongjian waike zazhi [Chinese Journal of Reparative and Reconstructive Surgery]. 1997;**11**(4):206-207

[60] Xu JJ, Wang Y. The effects of amniotic membrane on corneal penetration of ofloxacin. [Zhonghua yan ke za zhi] Chinese Journal of Ophthalmology. 2006;**42**(7):624-627

[61] Socci RR, Tachado SD, Aronstam RS, Reinach PS. Characterization of the muscarinic receptor subtypes in the bovine corneal epithelial cells. Journal of Ocular Pharmacology and Therapeutics. 1996;**12**(3):259-269. DOI: 10.1089/jop.1996.12.259

[62] Yang H, Sun X, Wang Z, Ning G, Zhang F, Kong J, et al. EGF stimulates growth by enhancing capacitative calcium entry in corneal epithelial cells. The Journal of Membrane Biology. 2003;**194**(1):47-58. DOI: 10.1007/s00232-003-2025-9

[63] Wu X, Yang H, Iserovich P, Fischbarg J, Reinach PS. Regulatory volume decrease by SV40-transformed rabbit corneal epithelial cells requires ryanodine-sensitive Ca^{2+}–induced Ca^{2+} release. The Journal of Membrane Biology. 1997;**158**(2):127-136. DOI: 10.1007/s002329900250

[64] Jeng BH, Meisler DM, Hollyfield JG, Connor JT, Aulak KS, Stuehr DJ. Nitric oxide generated by corneas in corneal storage media. Cornea. 2002;**21**(4):410-414. DOI: 10.1097/00003226-200205000-00015

[65] Meyer-Rochow VB, Mishra M. Structure and putative function of dark- and light-adapted as well as UV-exposed eyes of the food store pest *Psyllipsocus ramburi* Selys-longchamps (Insecta: Psocoptera: Psyllipsocidae). Journal of Insect Physiology. 2007;**53**(2):157-169. DOI: 10.1016/j.jinsphys.2006.11.002

[66] Lin MC, Jan CR. The anti-anginal drug fendiline elevates cytosolic Ca^{2+} in rabbit corneal epithelial cells. Life Sciences. 2002;**71**(9):1071-1079. DOI: 10.1016/s0024-3205(02)01797-6

[67] Ma D, Mellon J, Niederkorn JY. Conditions affecting enhanced corneal allograft survival by oral immunization. Investigative Ophthalmology & Visual Science. 1998;**39**(10):1835-1846

[68] Tao W, Wu X, Liou GI, Abney TO, Reinach PS. Endothelin receptor-mediated Ca^{2+} signaling and isoform expression in bovine corneal epithelial cells. Investigative Ophthalmology & Visual Science. 1997;**38**(1):130-141

[69] Lambiase A, Manni L, Bonini S, Rama P, Micera A, Aloe L. Nerve growth factor promotes corneal healing: Structural, biochemical, and molecular analyses of rat and human corneas. Investigative Ophthalmology & Visual Science. 2000;**41**(5):1063-1069

[70] Zhang F, Wen Q, Mergler S, Yang H, Wang Z, Bildin VN, et al. PKC isoform-specific enhancement of capacitative calcium entry in human corneal epithelial cells. Investigative

Ophthalmology & Visual Science. 2006;**47**(9):3989-4000. DOI: 10.1167/iovs.06-0253

[71] Xie Q, Zhang Y, Cai Sun X, Zhai C, Bonanno JA. Expression and functional evaluation of transient receptor potential channel 4 in bovine corneal endothelial cells. Experimental Eye Research. 2005;**81**(1):5-14. DOI: 10.1016/j.exer.2005.01.003

[72] Peterson RC, Wolffsohn JS. The effect of digital image resolution and compression on anterior eye imaging. The British Journal of Ophthalmology. 2005;**89**(7):828-830. DOI: 10.1136/bjo.2004.062240

[73] Xie Q, Zhang Y, Zhai C, Bonanno JA. Calcium influx factor from cytochrome P-450 metabolism and secretion-like coupling mechanisms for capacitative calcium entry in corneal endothelial cells. The Journal of Biological Chemistry. 2002;**277**(19):16559-16566. DOI: 10.1074/jbc.M109518200

[74] Yang H, Mergler S, Sun X, Wang Z, Lu L, Bonanno JA, et al. TRPC4 knockdown suppresses epidermal growth factor-induced store-operated channel activation and growth in human corneal epithelial cells. The Journal of Biological Chemistry. 2005;**280**(37):32230-32237. DOI: 10.1074/jbc.M504553200

[75] Petroll WM, Cavanagh HD, Lemp MA, Andrews PM, Jester JV. Digital image acquisition in in vivo confocal microscopy. Journal of Microscopy. 1992;**165**(Pt 1):61-69. DOI: 10.1111/j.1365-2818.1992.tb04305.x

[76] Greenwald R. The handling of corneal donor tissue before penetrating keratoplasty. Journal of Ophthalmic Nursing & Technology. 1994;**13**(1):21-22

[77] Socci R, Chu A, Reinach P, Meszaros LG. In situ Ca^{2+}-induced Ca^{2+} release from a ryanodine-sensitive intracellular Ca^{2+} store in corneal epithelial cells. Comparative Biochemistry and Physiology B, Comparative Biochemistry. 1993;**106**(4):793-797. DOI: 10.1016/0305-0491(93)90032-z

[78] Johnstone EW, Williams KA, Lovric VA, Lubeck D, Barras CW, Coster DJ. Cryopreservation of rabbit and cat corneas at −18 to −24 degrees C. Cornea. 1992;**11**(3):211-220

[79] Nishida T, Ueda A, Otori T, Fujita H. Long-term storage of endocytosed latex beads in keratocytes in vivo. Cornea. 1991;**10**(6):532-535. DOI: 10.1097/00003226-199111000-00012

[80] Elias KA, Weiner RI, Mellon SH. Effect of extracellular matrix on prolactin secretion and mRNA accumulation in GH3 cells. DNA and Cell Biology. 1990;**9**(5):369-375. DOI: 10.1089/dna.1990.9.369

[81] Roth HW. Deformation of the central corneal surface following use of hard and soft contact lenses. Fortschritte der Ophthalmologie: Zeitschrift der Deutschen Ophthalmologischen Gesellschaft. 1989;**86**(3):185-188

[82] Savion N, Disatnik MH, Nevo Z. Murine macrophage heparanase: Inhibition and comparison with metastatic tumor cells. Journal of Cellular Physiology. 1987;**130**(1):77-84. DOI: 10.1002/jcp.1041300112

[83] Bethea CL, Kozak SL. Effect of extracellular matrix on PC 12 cell shape and dopamine processing. Molecular and Cellular Endocrinology. 1984;**37**(3):319-329. DOI: 10.1016/0303-7207(84)90102-3

[84] Bethea CL, Kozak SL. Further characterization of substratum influence on PC12 cell shape and

dopamine processing. Molecular and Cellular Endocrinology. 1985;**42**(1):59-72. DOI: 10.1016/0303-7207(85)90007-3

[85] Bethea CL. Glucocorticoid stimulation of dopamine production in PC12 cells on extracellular matrix and plastic. Molecular and Cellular Endocrinology. 1987;**50**(3):211-222. DOI: 10.1016/0303-7207(87)90019-0

[86] Rich LF. A technique for preparing corneal lamellar donor tissue using simplified keratomileusis. Ophthalmic Surgery. 1980;**11**(9):606-608

[87] Vlodavsky I, Folkman J, Sullivan R, Fridman R, Ishai-Michaeli R, Sasse J, et al. Endothelial cell-derived basic fibroblast growth factor: synthesis and deposition into subendothelial extracellular matrix. Proceedings of the National Academy of Sciences of the United States of America. 1987;**84**(8):2292-2296. DOI: 10.1073/pnas.84.8.2292

[88] Ma DH, Chen HC, Ma KS, Lai JY, Yang U, Yeh LK, et al. Preservation of human limbal epithelial progenitor cells on carbodiimide cross-linked amniotic membrane via integrin-linked kinase-mediated Wnt activation. Acta Biomaterialia. 2016;**31**:144-155. DOI: 10.1016/j.actbio.2015.11.042

[89] McCarey BE, Kaufman HE. Improved corneal storage. Investigative Ophthalmology. 1974;**13**(3):165-173

[90] Lindstrom RL, Kaufman HE, Skelnik DL, Laing RA, Lass JH, Musch DC, et al. Optisol corneal storage medium. American Journal of Ophthalmology. 1992;**114**(3):345-356. DOI: 10.1016/s0002-9394(14)71803-3

[91] Ehlers H, Ehlers N, Hjortdal JO. Corneal transplantation with donor tissue kept in organ culture for 7 weeks. Acta Ophthalmologica Scandinavica. 1999;**77**(3):277-278. DOI: 10.1034/j.1600-0420.1999.770306.x

[92] Basu PK, Hasany S. Autolysis of the cornea of stored human donor eyes. Canadian Journal of Ophthalmology: Journal canadien d'ophtalmologie. 1974;**9**(2):229-235

[93] Pegg DE. Principles of cryopreservation. Methods in Molecular Biology (Clifton, NJ). 2007;**368**:39-57. DOI: 10.1007/978-1-59745-362-2_3

[94] Engineering challenges in tissue preservation. Cell Preservation Technology. 2004;**2**(2):91-112. DOI: 10.1089/153834404774101945

[95] Kim SY, Yang J, Lee YC. Safety of moxifloxacin and voriconazole in corneal storage media on porcine corneal endothelial cells. Journal of Ocular Pharmacology and Therapeutics: The Official Journal of the Association for Ocular Pharmacology and Therapeutics. 2010;**26**(4):315-318. DOI: 10.1089/jop.2009.0136

[96] Heindl LM, Riss S, Adler W, Bucher F, Hos D, Cursiefen C. Split cornea transplantation: Relationship between storage time of split donor tissue and outcome. Ophthalmology. 2013;**120**(5):899-907. DOI: 10.1016/j.ophtha.2012.11.012

[97] Lopez-Garcia JS, Garcia-Lozano I, Rivas L, Ramirez N, Mendez MT, Raposo R. Stability of growth factors in autologous serum eyedrops after long-term storage. Current Eye Research. 2016;**41**(3):292-298. DOI: 10.3109/02713683.2015.1016180

[98] Borderie VM, Lopez M, Lombet A, Carvajal-Gonzalez S, Cywiner C, Laroche L. Cryopreservation and culture of human corneal keratocytes. Investigative Ophthalmology & Visual Science. 1998;**39**(8):1511-1519

[99] Fischer KR, Opitz A, Boeck M, Geerling G. Stability of serum eye drops after storage of 6 months. Cornea.

2012;**31**(11):1313-1318. DOI: 10.1097/ICO.0b013e3182542085

[100] Lai JY, Wang PR, Luo LJ, Chen ST. Stabilization of collagen nanofibers with L-lysine improves the ability of carbodiimide cross-linked amniotic membranes to preserve limbal epithelial progenitor cells. International Journal of Nanomedicine. 2014;**9**:5117-5130. DOI: 10.2147/ijn.s69689

[101] Guindolet D, Crouzet E, He Z, Herbepin P, Jumelle C, Perrache C, et al. Storage of porcine cornea in an innovative bioreactor. Investigative Ophthalmology & Visual Science. 2017;**58**(13):5907-5917. DOI: 10.1167/iovs.17-22218

[102] Sabater AL, Guarnieri A, Espana EM, Li W, Prosper F, Moreno-Montanes J. Strategies of human corneal endothelial tissue regeneration. Regenerative Medicine. 2013;**8**(2):183-195. DOI: 10.2217/rme.13.11

[103] Patel D, Tandon R, Ganger A, Vij A, Lalwani S, Kumar A. Study of death to preservation time and its impact on utilisation of donor corneas. Tropical Doctor. 2017;**47**(4):365-370. DOI: 10.1177/0049475517713406

[104] Kim KY, Jung JW, Kim EK, Seo KY, Kim TI. Tectonic lamellar keratoplasty using cryopreserved cornea in a large descemetocele. Yonsei Medical Journal. 2016;**57**(1):269-271. DOI: 10.3349/ymj.2016.57.1.269

[105] Duncan K, Parker J, Hoover C, Jeng BH. The effect of light exposure on the efficacy and safety of amphotericin B in corneal storage media. JAMA Ophthalmology. 2016;**134**(4):432-436. DOI: 10.1001/jamaophthalmol.2016.0008

[106] Cunningham WJ, Moffatt SL, Brookes NH, Twohill HC, Pendergrast DG, Stewart JM, et al. The New Zealand National Eye Bank study: Trends in the acquisition and storage of corneal tissue over the decade 2000 to 2009. Cornea. 2012;**31**(5):538-545. DOI: 10.1097/ICO.0b013e318222c3f2

[107] Lee JK, Ryu YH, Ahn JI, Kim MK, Lee TS, Kim JC. The effect of lyophilization on graft acceptance in experimental xenotransplantation using porcine cornea. Artificial Organs. 2010;**34**(1):37-45. DOI: 10.1111/j.1525-1594.2009.00789.x

[108] Mauger T, Quartetti E. The effects of the presence of the corneal epithelium and supplemental hydrocortisone on beta-glucuronidase levels with corneal preservation. Clinical Ophthalmology. 2011;**5**:415-417. DOI: 10.2147/opth.s17228

[109] Daoud YJ, Smith R, Smith T, Akpek EK, Ward DE, Stark WJ. The intraoperative impression and postoperative outcomes of gamma-irradiated corneas in corneal and glaucoma patch surgery. Cornea. 2011;**30**(12):1387-1391. DOI: 10.1097/ICO.0b013e31821c9c09

[110] Patel HY, Ormonde S, Brookes NH, Moffatt SL, Sherwin T, Pendergrast DG, et al. The New Zealand National Eye Bank: Survival and visual outcome 1 year after penetrating keratoplasty. Cornea. 2011;**30**(7):760-764. DOI: 10.1097/ICO.0b013e3182014668

[111] Feilmeier MR, Tabin GC, Williams L, Oliva M. The use of glycerol-preserved corneas in the developing world. Middle East African Journal of Ophthalmology. 2010;**17**(1):38-43. DOI: 10.4103/0974-9233.61215

[112] Yang JW, Lin HC, Hsiao CH, Chen PY. Therapeutic penetrating keratoplasty in severe infective keratitis using glycerol-preserved donor corneas. Cornea. 2012;**31**(10):1103-1106. DOI: 10.1097/ICO.0b013e31821c9ba2

[113] Fabiano A, Bizzarri R, Zambito Y. Thermosensitive hydrogel based on chitosan and its derivatives containing medicated nanoparticles for transcorneal administration of 5-fluorouracil. International Journal of Nanomedicine. 2017;**12**:633-643. DOI: 10.2147/ijn.s121642

[114] Gaum L, Reynolds I, Jones MN, Clarkson AJ, Gillan HL, Kaye SB. Tissue and corneal donation and transplantation in the UK. British Journal of Anaesthesia. 2012;**108** (Suppl 1):i43-i47. DOI: 10.1093/bja/aer398

[115] Costa D, Leiva M, Naranjo C, Rios J, Pena MT. Cryopreservation (−20°C) of feline corneoscleral tissue: Histologic, microbiologic, and ultrastructural study. Veterinary Ophthalmology. 2016;**19**(Suppl 1):97-104. DOI: 10.1111/vop.12393

[116] Hashimoto Y, Hattori S, Sasaki S, Honda T, Kimura T, Funamoto S, et al. Ultrastructural analysis of the decellularized cornea after interlamellar keratoplasty and microkeratome-assisted anterior lamellar keratoplasty in a rabbit model. Scientific Reports. 2016;**6**:27734. DOI: 10.1038/srep27734

[117] Tandon R, Singh A, Gupta N, Vanathi M, Gupta V. Upgradation and modernization of eye banking services: Integrating tradition with innovative policies and current best practices. Indian Journal of Ophthalmology. 2017;**65**(2):109-115. DOI: 10.4103/ijo.IJO_862_16

[118] Gupta N, Upadhyay P. Use of glycerol-preserved corneas for corneal transplants. Indian Journal of Ophthalmology. 2017;**65**(7):569-573. DOI: 10.4103/ijo.IJO_56_17

[119] Liu Z, Zhou Q, Zhu J, Xiao J, Wan P, Zhou C, et al. Using genipin-crosslinked acellular porcine corneal stroma for cosmetic corneal lens implants. Biomaterials. 2012;**33**(30):7336-7346. DOI: 10.1016/j.biomaterials.2012.06.080

[120] Miller TD, Maxwell AJ, Lindquist TD, Requard J 3rd. Validation of cooling effect of insulated containers for the shipment of corneal tissue and recommendations for transport. Cornea. 2013;**32**(1):63-69. DOI: 10.1097/ICO.0b013e3182685c8e

[121] Rijneveld WJ, Wolff R, Volker-Dieben HJ, Pels E. Validation of tissue quality parameters for donor corneas designated for emergency use in preservation of the globe. Cornea. 2010;**29**(2):128-132. DOI: 10.1097/ICO.0b013e3181ac07bc

[122] Bi YL, Bock F, Zhou Q, Cursiefen C. Central corneal epithelium self-healing after ring-shaped glycerin-cryopreserved lamellar keratoplasty in Terrien marginal degeneration. International Journal of Ophthalmology. 2013;**6**(2):251-252. DOI: 10.3980/j.issn.2222-3959.2013.02.27

[123] Liu X, Shen JH, Zhou Q, Liu ZX, Tang SF, Chen RR, et al. Personalised lamellar keratoplasty and keratopigmentation in Asian corneal leucoma patients. International Journal of Clinical and Experimental Medicine. 2015;**8**(6):9446-9453

[124] Liu X, Zhou Q, Huang X, Liu Z, Bi Y. Clinical evaluation of deep anterior lamellar keratoplasty using glycerol-cryopreserved corneal tissues for refractory herpetic stromal keratitis: An observational study. Medicine (Baltimore). 2016;**95**(39):e4892. DOI: 10.1097/MD.0000000000004892

Section 4

Dynamics of Water Content in Plant Tissues During Cooling and Heating

Chapter 8

Methods of Thermal Analysis as a Tool to Develop Cryopreservation Protocols of Vegetatively Propagated Crops

Stacy D. Hammond, Miloš Faltus and Jiří Zámečník

Abstract

Cryopreservation is considered to be a reliable biotechnological tool for the long-term conservation of vegetatively propagated plant germplasm. The technique is based on freezing plant tissues at an ultralow temperature. However, high water content in plant tissue can result in injury during the cooling and thawing processes. Water behavior in the process of cryopreservation can be assessed by the use of thermal analysis method. This chapter demonstrates how the use of heat flux-type differential scanning calorimetry (DSC) thermal analysis methods such as standard DSC, temperature-modulated DSC (TMDSC), and quasi-isothermal temperature-modulated DSC (QITMDSC) can be used to assess the amount of freezable water and verify if the tissue being used has reached glass transition as well as analyzing the thermal events during cooling and freezing to reduce crystallization and damage by frost. Here, you can find a guide on how these thermal analysis methods can be applied, through concrete examples of each method and their use in the development of a more reliable and precise cryopreservation protocol for vegetatively propagated plant species.

Keywords: differential scanning calorimetry, freezable water, glass transition, heat flux DSC, ice crystallization, quasi-isothermal temperature-modulated differential scanning calorimetry, temperature-modulated differential scanning calorimetry

1. Introduction

Cryopreservation is emerging as one of the most promising techniques for the long-term conservation of plant germplasm; it is also considered as an alternative method to safeguard plant species preserved by conventional conservation practices such as in vitro techniques and ex situ methods, i.e., seed and field collections, including vegetatively propagated crop species.

This conservation technique is based on the freezing of tissues (e.g., organs and shoots tips) from in vitro plantlets or field collections at extremely low temperatures, with the aim of reducing metabolic activity while at the same time maintaining the vitality of the tissue. The main principle of the method is based on cooling plant tissues at an ultralow temperature, usually by the use of liquid nitrogen (LN) which has a constant temperature of −196°C. The frozen tissue after the desired period in ultralow temperature after rewarming is able to regenerate to healthy plants. However, this method is not as straightforward as it seems; freezing plant tissue can result in intracellular ice nucleation and subsequent ice crystallization leading to cell damage during cooling and/or rewarming of the samples.

Low water content minimizes ice crystallization. Hence, the objective status for cryopreservation is to reduce the water content in the plant tissue. By this procedure it obtains a glassy state to avoid the formation of lethal intracellular ice crystals in order to obtain optimal recovery and regeneration after cryopreservation. This can be achieved by the induction of vitrification status which is a glass induction by dehydration, the addition of cryoprotectants, and a very fast decrease in temperature.

Methods based on dehydration include osmotic dehydration, air desiccation, and freeze dehydration [1–3]. The state of the water in the plant during cooling/warming cycles is shown in **Figure 1** with highlighted thermally colored and its possible detection and quantification by DSC techniques.

Various plant vitrification solutions (PVS) are used for osmotic dehydration; these are usually labeled with a number according to the specific mixture of basic cryoprotectants and their concentrations. The main PVS are Luyet [4], Fahy [5], Steponkus [6], PVS1 [7, 8], PVS2 [9], PVS3, PVS4, PVS5 [10], VS6 [11], PVSL [12], and VSL [9]. These are composed of different concentrations and combinations of four main cryoprotectants: dimethyl sulfoxide, sucrose, glycerol, and ethylene glycol.

Each cryopreservation protocol is species- and genotype-specific and needs to be optimized accordingly. Hence, the standardization of the protocol is necessary to ensure and facilitate an effective cryo-storage of the plant germplasm. To achieve this, it is necessary to determine the optimal water content during dehydration, freezing in LN and thawing to avoid ice crystallization. Additionally, assessing the amount of freezable water and verifying if the tissue being used has reached glass transition as well as analyzing the thermal events during cooling and freezing are key factors in order to develop and standardize a reliable protocol.

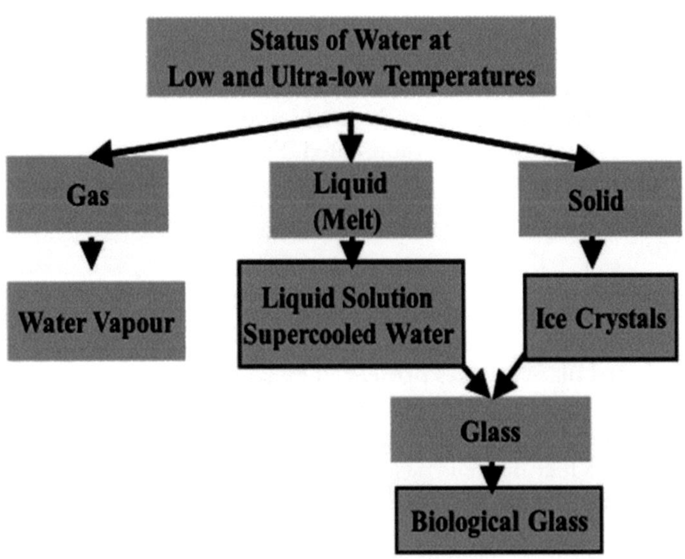

Figure 1.
State of water in biological objects for cryopreservation. Those highlighted in a black frame are possible to quantify by DSC.

This can be determined by the use of thermal analysis such as differential scanning calorimetry (DSC).

The use of DSC for checking the content of freezable water (or first-order water transitions freezing and melting) is applied for the cryopreservation of shoot tips [13–15], dormant buds/winter-hardy buds [16–18], pollen [19–21], and seeds [15, 22–24]. The development of cryopreservation protocols using DSC has been used for a number of plant species, i.e., potato [25], *Ullucus tuberosus* Cal. [26], *Oxalis tuberosa* Mol. [27], *Allium* species [19, 21, 28, 29], apple [30], and *Actinidia* spp. [31], among others. However, the method is still rarely used in cryobiology due to the lack of awareness of its use, application, and potential in this field. This chapter intends to broaden the knowledge and demonstrate the use of thermal analysis methods with heat flux-type DSC and their applicability in cryopreservation through a series of research carried out at the Crop Research Institute in Prague, Czech Republic. The results presented in this chapter represent original unpublished work archived at the authors' institution.

2. Differential scanning calorimetry

DSC is a thermal method that can be used to measure and determine the phase transitions for cryopreservation. Using a function of time and temperature, the main principle of this method is to measure the temperatures and heat flows associated with thermal transitions in plant material providing information on the endothermic or exothermic events or changes in heat capacity. This information can be used to determine the glass transition, the temperature of ice nucleation, melting, boiling, crystallization time, and kinetic reaction which are the most important characteristics useful for cryopreservation [16].

This method is based on applying a regulated decrease and increase in temperature while measuring the heat flow and temperature corresponding to the tested sample. The heat flux-type DSC measures differences of temperature between reference and sample and recalculates the differential heat flux. The most common cooling/heating rate of the sample is 10°C min^{-1}. There are three different temperature modifications of the heat flux-type DSC: standard DSC gives a basic information about exothermic and endothermic signals and about glass transition presence, temperature-modulated DSC (TMDSC) separates kinetic and thermodynamic events in tested samples, and quasi-isothermal temperature-modulated DSC (QITMDSC) determines the exact measurement of heat capacity at equilibrium conditions. The principle, application, and use of each of these DSC modifications will be demonstrated within this chapter through a series of experiments carried out on vegetatively propagated species and cryoprotectants.

Thermal characteristics as melting point temperature, glass transition temperature, and proportion of freezable water content (percent of the fresh weight) were determined using a differential scanning calorimeters TA2920 (TA Instruments, USA) with refrigerated cooling system (RCS) in the temperature range from −60 to +20°C and Q2000 (TA Instruments, USA) with RCS or liquid nitrogen cooling system (LNCS) in the temperature range from −90 to +20°C or from −140 to +20°C, respectively. The cooling and heating rates were 10°C/min. or 1°C/min. For standard DSC or TMDSC, respectively. Temperature modulation was performed at 1°C amplitude of modulation and 60-second period. Aluminum, hermetically sealed pans were used for all samples to avoid water evaporation before and during measurement cycles. Sample size ranged from 5 to 20 mg. The purge gas was either nitrogen or helium.

3. Standard DSC measurement and result analyses

Standard DSC is a thermal analysis method used to carry out a total heat flow measurement of the sample in order to determine phase transitions such as crystallization, melting, and glass transition (**Figure 2**).

In terms of cryopreservation of plant tissue, this method of thermal analysis can help to improve the development of a more reliable protocol. It can be applied in the phase of optimization of the sample's preculture, acclimation, or dehydration to decrease the amount of freezable water content within the sample and in the risk of sample injury (**Figure 3**). The quality of cryo-sample preparation can be checked and modified to avoid the presence of cold crystallization in the sample during its thawing after cryo- storage for its recovery (**Figure 4**). This method can also be used to assess the T_g (glass transition temperature) and to determine safe storage temperatures for the samples.

3.1 Freezable water content determination

Freezable water content is probably the most important parameter which influences the injury of plant tissue during cryopreservation using cryoprotocol avoiding controlled freeze dehydration. The proper decrease in freezable water content is hence prerequisite for successful cryopreservation. Standard DSC method measures the endothermic and exothermic signal related to water freezing and thawing. Based on the knowledge of heat of fusion specific for water at 0°C (334 J/g/K) and total amount of energy used for the phase transition of water melting (ice crystals thawing), the total amount of frozen water can be calculated, and based on this finding, the optimal time of PVS3 treatment can be adjusted. Freezable water was detected in garlic shoot tips during PVS3 dehydration for a period up to 90 min (**Figure 5a**). After ice nucleation this water fraction can cause cell damage by ice crystal growth. Intracellular ice formation is in any case lethal for the plant cell. There was no freezable water in plants after 90 min of dehydration. The appropriate level of dehydration by PVS3 has to be

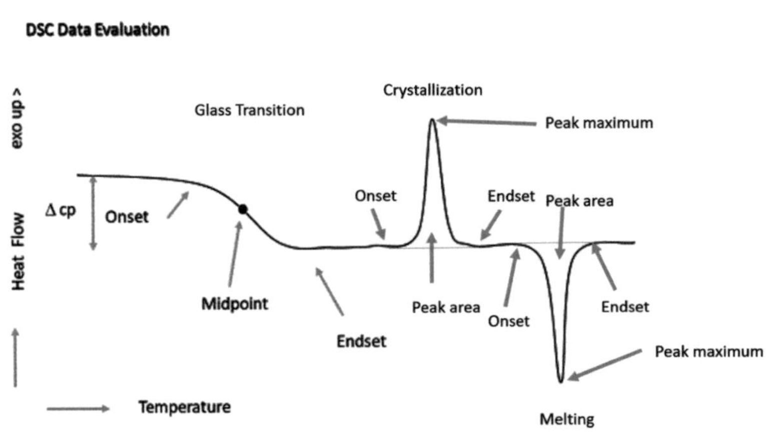

Figure 2.
Thermogram from differential scanning calorimeter during warming. The glass transition is characterized by the beginning of the heat flow change (onset), midpoint (in some cases inflection point), and finish of heat flow change (endset), characterized by change of heat capacity (ΔCp). Crystallization of supercooled water is an exothermic reaction, followed by the thawing of crystalized water. These events can be characterized by onset, midpoint, and endset temperatures and by event heat flow change. Based on analysis of the exothermic or endothermic events, the freezable water content can be calculated also in dehydrated samples. Additionally, this thermal analysis method is a powerful tool in the assessment of the sample's water behavior during cooling and warming; this includes supercooling, freezing, glass transition, cold crystallization, and melting which influence the sample's cryopreservation success.

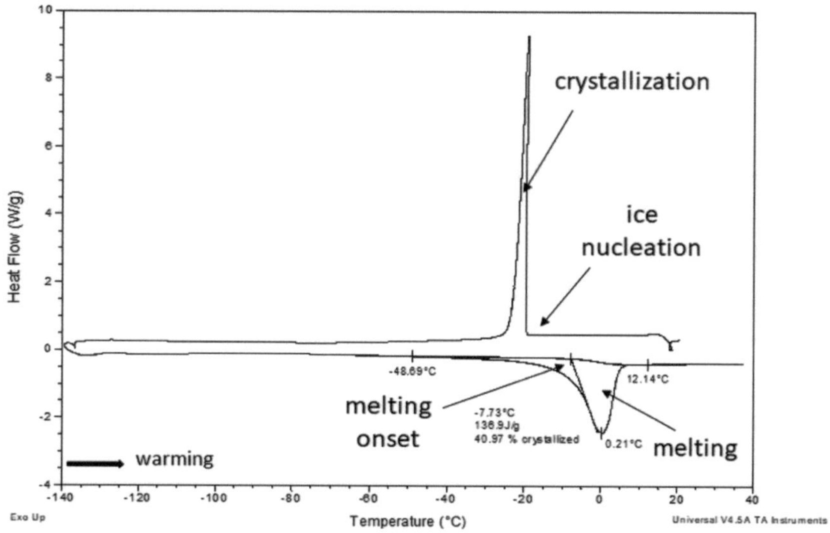

Figure 3.
Area of the melting peaks represents the heat of fusion needed for melting of the sample, related to the amount of frozen water in potato shoot tips (cv. "Arnika") after 1 h air dehydration (Q2000 + LNCA).

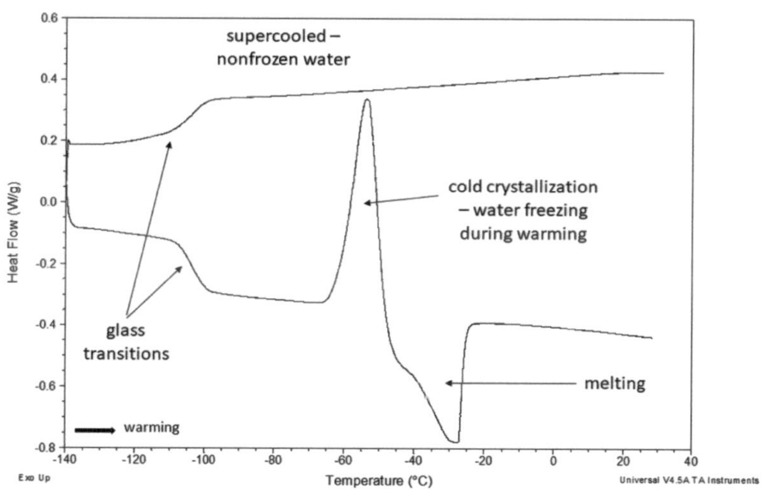

Figure 4.
Cold crystallization during the warming cycle in potato shoot tips (cv. "Arnika") in diluted PVS3 solution (Q2000 + LNCA).

controlled due to the decrease of survival and regeneration of shoot tips by excessive dehydration (**Figure 5b**). For successful cryopreservation with high regeneration rate, samples in narrow dehydration window of PVS3 must be used: with neglected or no freezable water (**Figure 5a**) and high regeneration rate due to dehydration (**Figure 5b**) after warming from ultralow temperatures.

3.2 Safe storage temperature determination

Safety and stability of plant sample stored at ultralow temperature are influenced by the temperature of glass transition. Throughout the period of cryopreservation, the sample has to be kept below the sample glass transition temperature.

In case of increase of storage above the sample glassy state temperature, the risk of sample damage by ice crystals occurs. Different cryopreservation methods resulted in different sample glass transition temperature. Thus, the selection of the cryopreservation method or cryoprotectant used can influence safe storage temperature. The most frequently used cryoprotectants for plant cryopreservation PVS2 and PVS3 are characterized by Tg of −114 and −90°C, respectively. When the 80%PVS3 solutions were used, the glass transition temperature was approximately −88°C in potato shoot tips dehydrated for 2 h (**Figure 6**).

Some cryoprotective solutions with a higher glass transition temperature can be used to increase the safe storage temperature and sample stability. Standard cryopreservation methods used in CRI is based on sucrose loading the shoot tips and subsequent air dehydration. This dehydration resulted in increased sucrose concentration and as a result increased in glass transition temperature. Thus, the glass transition temperature of potato shoot tips increased to approximately −23°C after 0.7 M sucrose loading and following air dehydration for 5 h (**Figure 7**).

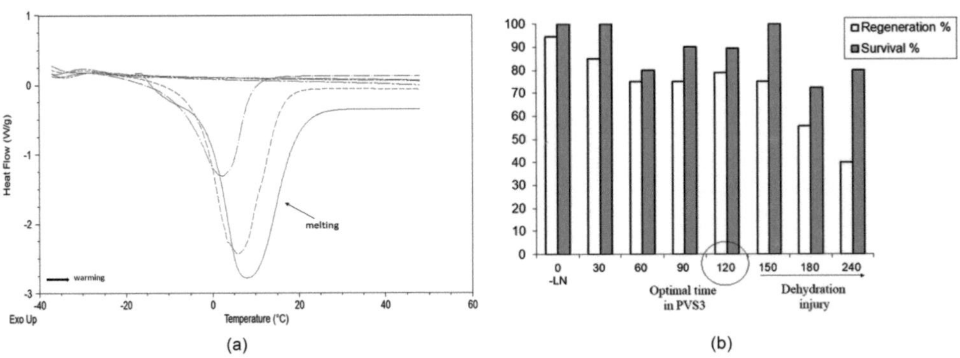

Figure 5.
(a) Frozen water content after different time of shoot tips of Allium sativum *"Djambul" in PVS3, after 30, 60, and 90 min (from largest to smallest peak). There is no detectable frozen water content in shoot tips for a longer time in PSV3 (TA 2920 + RCS). (b) Survival and regeneration after different time of shoot tips of* Allium sativum *"Djambul" in PVS3, after 30, 60, and 90 min (in this time interval, there was no regeneration after liquid nitrogen immersion, because of lethal injury from frozen water). The optimal time is 120 min in PVS3 because from this time the survival and regeneration rate decreased because of dehydration injury*

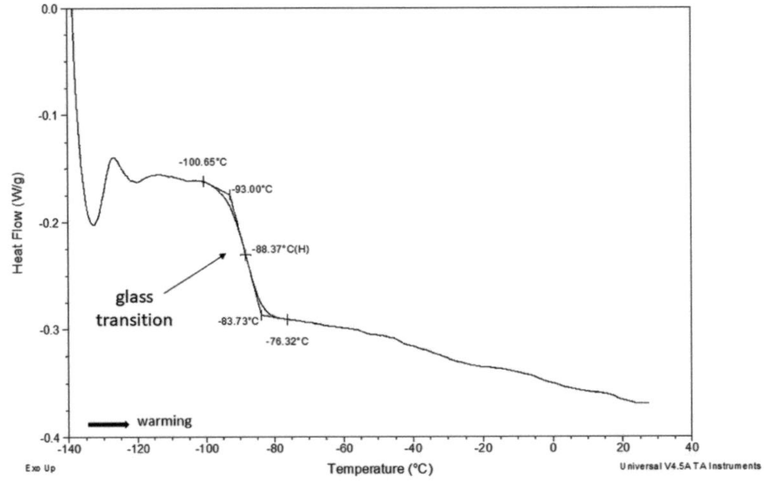

Figure 6.
Glass transition of potato shoot tips (cv. "Désirée") after 2 h dehydration in 80%PVS3 (Q2000 + LNCA).

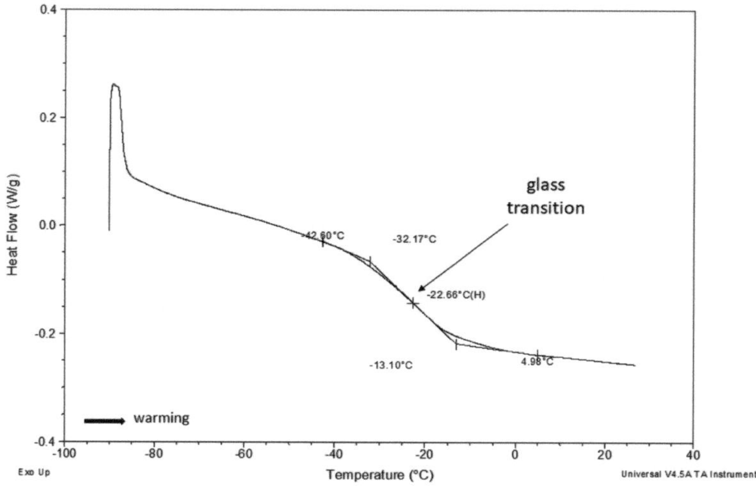

Figure 7.
Glass transition of potato shoot tips (cv. "Désirée") loaded in 2 M sucrose solution and subsequently dehydrated by air for 5 h (Q2000 + RCS).

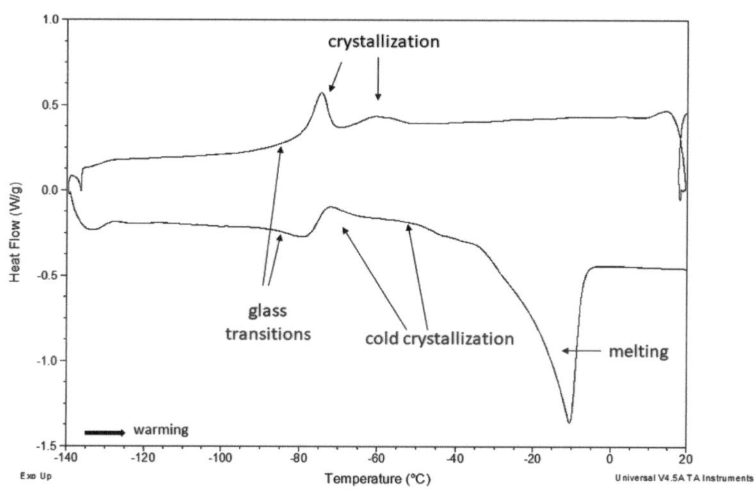

Figure 8.
Complex thermogram of 60% sucrose solutions including thermodynamic (glass transition, melting) and kinetic (crystallization) events (Q2000 + LNCA).

3.3 Benefits and limits of standard DSC measurements

Though this method of thermal analysis is easy and simple to use (most standard programs use simple methods with cooling/warming rates of 10°C min^{-1}) and provides a relatively fast real-time measurement of thermal characteristics in the course of dehydration of the sample (e.g., 30-min DSC program allows to perform measurements in every 30 min of dehydration), there are some limitations. One of these limitations is the measurement of the total heat flow signals which include both thermodynamic and kinetic signals; hence, the evaluation of thermodynamic signals may not be exact. Additionally, kinetic events like relaxation or crystallization make thermogram evaluation and event identification difficult and can mask the thermodynamic component of the total heat flow signal (**Figure 8**). The overlapping signals may not be identified. In such cases, there is the need to carry out a more detailed analysis of the specific thermal events by the use of TMDSC or QITMDSC.

4. Temperature-modulated DSC

TMDSC is the thermal analysis method (**Figure 9**) used to separate the total heat flow signal into thermodynamic and kinetic components. Hence, through this method, thermodynamic events such as melting, glass transition temperatures, and kinetic events such as crystallization and relaxations can be analyzed separately. Additionally, the freezable water content and the water behavior in the sample during cooling/warming cycles, like supercooling, freezing, glass transition, cold crystallization, and melting can be determined. This method also determines the Cp (specific heat capacity) and allows to specify its reversible Cp (thermodynamic) component.

This method of thermal analysis can help to improve the development of a cryopreservation protocol by improving the quality of the cryo-sample preparation. The method can identify kinetic events, which mask the thermodynamic events and which reveal unstable sample conditions. Due to the identification of the thermodynamic component of the heat flow signal by this method, the exact determination of freezable water content and the Tg in the sample is possible.

4.1 Separation of thermodynamic and kinetic events to identify the stability of plant material

This method of thermal analysis can help to improve the development of a cryopreservation protocol by improving the quality of the cryo-sample preparation. The method can identify kinetic events, which mask the thermodynamic events and which reveal unstable sample conditions. Due to the identification of the thermodynamic component of the heat flow signal by this method, the exact determination of freezable water content and the Tg in the sample is possible.

The total heat flow signal includes both thermodynamic and kinetic signals. That is why exact analysis in complex thermogram is difficult (**Figure 10**). Total heat flow showed some changes in the region from −61 to −12°C in 64% sucrose solution (from left to right): flat-shaped curve, high s-shaped curve, and peak. It is difficult to solve if the second event is a glass transition of melting. The TMDSC separated total heat flow into reversing and non-reversing signals. The only two

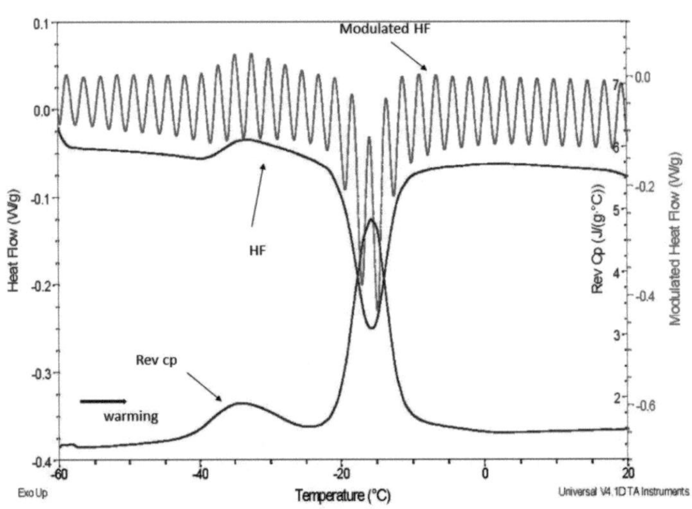

Figure 9.
The example of TMDSC result of thawing thermal event in Allium shoot tips. Note the reversing heat capacity (rev Cp) opposite to heat flux (TA 2920 + RCS).

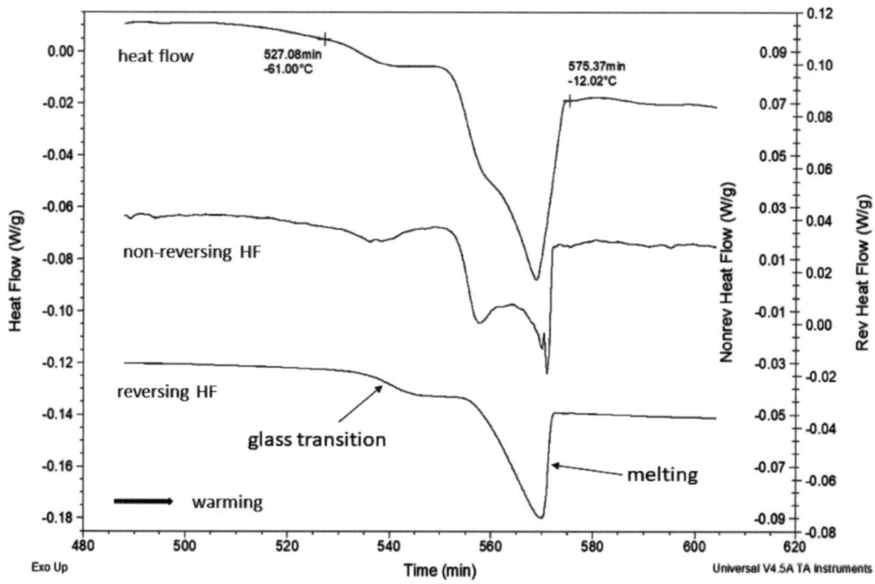

Figure 10.
Separation of total heat flow (HF) signal (upper) to non-reversing kinetic signal (middle) and reversing thermodynamic signal (bottom) in 64% sucrose solution (Q2000 + LNCA).

events were detected on reversing curve: glass transition and melting. The second event detected by standard DSC represents just kinetic event related to the crystallization of free water which is released before from glassy state during warming.

4.2 State of matter determination by a direct Cp measurement

The TMDSC can be also used for direct measurement of specific heat capacity (Cp). It changes with the change of the state of matter. The liquid state is characterized by a higher value of Cp than solid (crystal or ice) states. Phase changes, glass transition, and melting are characterized by a specific shape of Cp curve. The glassy state is characterized by an S-shaped curve of Cp baseline; on the other hand, melting is characterized by a peak (**Figure 11**).

Proper sample preparation for its successful cryopreservation is characterized by low-temperature glass transition curve but without melting peak presence. A decrease in the melting peak on the Cp curve is therefore necessary during sample preparation. An appropriate way of dehydration has to be applied to avoid plant tissue injury by excessive dehydration. Tolerance to dehydration is influenced by species, genotype, and by plant tissue or cells' physiological status. Pollen, besides orthodox seeds, is a plant material that can be dehydrated to very low water content without decrease of plant material vitality. Dehydrated pollen is characterized by the absence of a melting peak but the presence of glass transition s-shaped curve (**Figure 12**). Due to very low water content, the glass transition temperature is at high temperature (−18°C), which strongly increases the safe storage temperature of the material during cryopreservation.

4.3 Benefits and limits of TMDSC

Though if properly applied and analyzed, the method can be advantageous because of its exact measurement of thermodynamic events due to the separation of thermodynamic and kinetic signals and can also be useful in the identification of the overlapping events.

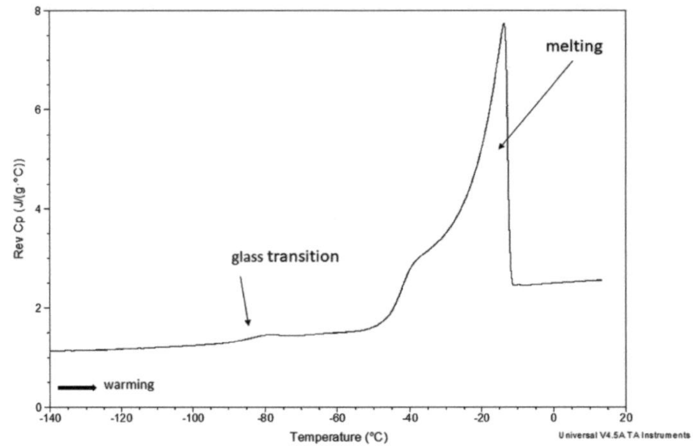

Figure 11.
Changes in reversible Cp during warming with detection of glass transition and melting in 20% sucrose solution by TM DSC (Q2000 + LNCA).

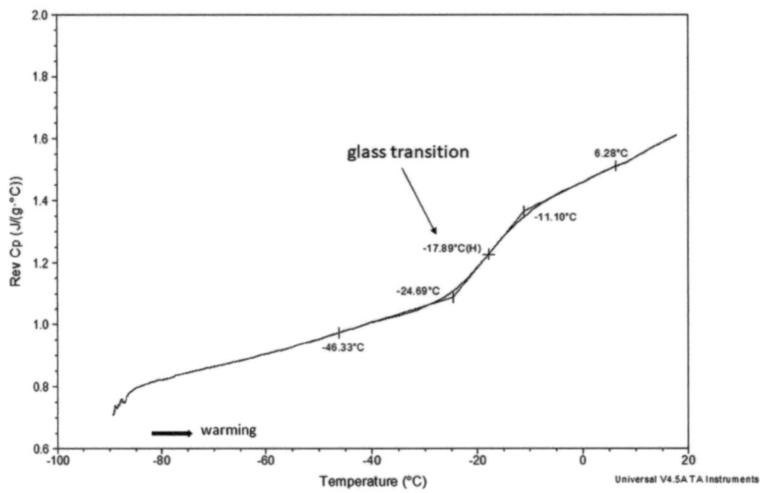

Figure 12.
Changes in reversible Cp during warming with detection of the glass transition in hop pollen (Q2000 + RCS).

TMDSC solved the limits of standard DSC method, but it has some limits too. It is time-consuming due to slower cooling/warming ramp rates. This can be partially solved by temperature modulation in narrower temperature range, which could be previously identified by a standard DSC measurement. TMDSC is sensitive to specific parameter measurement settings, which may result in failure to determine the events due to inadequate parameters of measure. Additionally, the interpretation of the results obtained can be difficult as it carries out a more detailed analysis which provides detailed results that are difficult to interpret (e.g., kinetic events).

5. Quasi-isothermal temperature-modulated DSC

QITMDSC is a specific variant of TMDSC. This method of thermal analysis is used for the direct measurement of complex Cp at given temperatures in

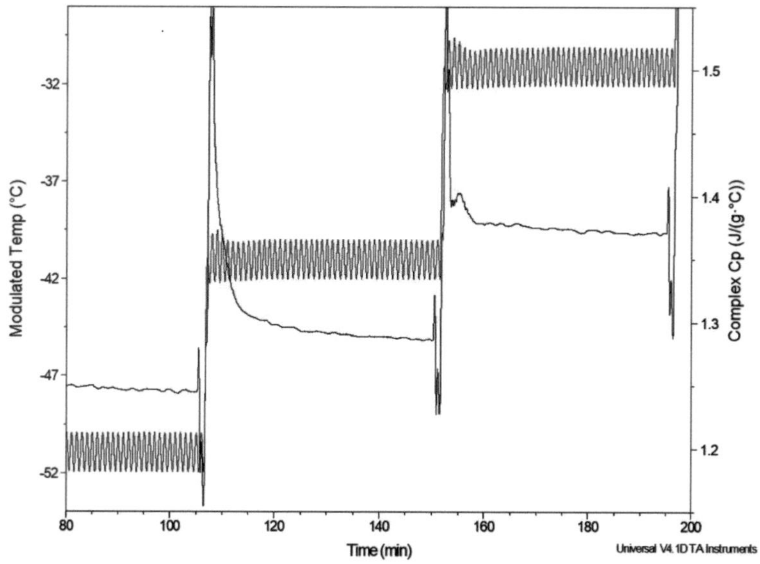

Figure 13.
Temperature modulation at a constant temperature until heat capacity equilibrates. After equilibration of heat capacity, the temperature jumps to higher temperature and modulation starts again (TA 2920 + RCS).

equilibrium conditions, at zero ramp cooling/warming rates (**Figure 13**). Hence, it is used to determine the Cp, more specifically, the thermodynamic events such as melting and glass transition.

5.1 The exact measurement of Cp under equilibrium conditions

The QITMDSC measures sample heat capacity (Cp) similarly like in the case of MDSC. That is why similar achievements can be obtained. This method can be used together with a method of matter relaxation during annealing at increasing temperature. The advantage of this method is the following measurement of the sample at stable conditions because all processes can be finished during modulation at equilibrium conditions (temperature). The glassy state is also characterized by an S-shaped curve of Cp baseline; on the other hand, melting is characterized by a peak (**Figure 14**).

5.2 Benefits and limits of TMDSC

This method is applied in the development of a cryopreservation protocol as it is useful in the determination of the state of matter and exact measurement of glass transition and melting. Additionally, it is adventitious in obtaining a more exact measurement of Cp as the Cp equilibrium conditions avoid the effects of kinetic events.

However, there are a few limitations in using QITMDSC as it is also very time-consuming, though this can be partially solved by analysis in limited temperature range, which was previously analyzed by a standard DSC or TMDSC analysis. The method is also sensitive to the parameters of measurement settings which may result in failure to determine the events due to inadequate parameters of measure. Nevertheless, as is the case with TMDSC if applied correctly, this method represents a useful tool when developing a cryopreservation protocol for vegetatively propagated plant species.

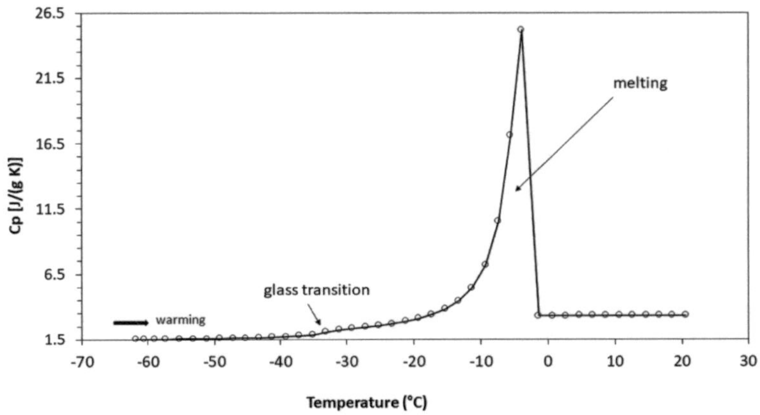

Figure 14.
Changes in Cp revealed the glass transition and the melting peak in 20% sucrose solution analyzed by QITMDSC after relaxation at −6°C (Q2000 + RCS).

6. Conclusion

Understanding the dynamics of water content in plant tissues during cooling and heating is crucial in developing a reliable cryopreservation protocol. Differential scanning calorimetry thermal analysis methods such as standard DSC, temperature-modulated DSC, and quasi-isothermal temperature-modulated DSC play a key role when cryopreserving plant material as they not only broaden the knowledge of thermal events but can also help to overcome the freezing injury during cryopreservation. The standard DSC method is recommended for routine work with known thermal properties of the sample and nonoverlapping thermal events. The MDSC method is recommended to use for plant samples with overlapping thermal events or for exact measurement of thermodynamic events. The hidden thermal events can be arisen and be separated by temperature modulation. The MDSC is also recommended to distinguish between reversing and non-reversing thermal events. The measurement of heat flow and heat capacity can be done in a single experiment. The QITMDSC is recommended to use for exact measurement of heat capacity in equilibrated conditions which can help a state of matter identification.

Acknowledgements

The authors would like to thank Mrs. Zdena Hybnerova and MS. Adrijana Cajkova for their technical help. This work is partially supported by projects QK1910277, QK1910476, and RO0418 from the Czech Ministry of Agriculture.

Author details

Stacy D. Hammond, Miloš Faltus* and Jiří Zámečník
Crop Research Institute, Prague, Czech Republic

*Address all correspondence to: faltus@vurv.cz

References

[1] Benson EB, Reed BM, Brennan RM, Clacher KA, Ross DA. Use of thermal analysis in the evaluation of cryopreservation protocols for *Ribes nigrum* L germplasm. Cryo-Letters. 1996;**17**(6):347-362. ISSN: 0143-2044

[2] Grospietsch M, Stodůlková E, Zámečník J. Effect of osmotic stress on the dehydration tolerance and cryopreservation of *Solanum tuberosum* shoot tips. Cryo-Letters. 1999;**20**(6): 339-346. ISSN: 0143-2044

[3] Hirai D, Sakai A. Cryopreservation of in vitro-grown meristems of potato (*Solanum tuberosum* L.) by encapsulation-vitrification. Potato Research. 1999;**42**(2):153-160. DOI: 10.1007/BF02358405

[4] Luyet B. On the mechanism of cellular death by high-pressure cytological modifications accompanying death in yeast. Comptes Rendus Hebdomadaires Des Seances De l Academie Des Sciences. 1937;**204**: 1506-1508. ISSN: 0001-4036

[5] Fahy GM. Cold shock injury is a significant factor in freezing-injury—A position against. Cryobiology. 1985;**22**(6):628-628. DOI: 10.1016/0011-2240(85)90111-7

[6] Langis R, Steponkus PL. Cryopreservation of rye protoplasts by vitrification. Plant Physiology. 1990;**92**(3):666-671. DOI: 10.1104/pp.92.3.666

[7] Uragami A, Sakai A, Nagai M, Takahashi T. Survival of cultured-cells and somatic embryos of *Asparagus officinalis* cryopreserved by vitrification. Plant Cell Reports. 1989;**8**(7):418-421. DOI: 10.1007/BF00270083

[8] Towill LE. Cryopreservation of isolated mint shoot tips by vitrification. Plant Cell Reports. 1990;**9**(4):178-180. DOI: 10.1007/BF00232174

[9] Suzuki M, Tandon P, Ishikawa M, Toyomasu T. Development of a new vitrification solution, VSL, and its application to the cryopreservation of gentian axillary buds. Plant Biotechnology Reports. 2008;**2**(2):123-131. DOI: 10.1007/s11816-008-0056-5

[10] Nishizawa S, Sakai A, Amano Y, Matsuzawa T. Cryopreservation of Asparagus (*Asparagus officinalis* L) embryogenic suspension cells and subsequent plant regeneration by vitrification. Plant Science. 1993;**91**(1):67-73. DOI: 10.1016/0168-9452(93)90189-7

[11] Liu HQ, Yu WG, Dai JX, Gong QH, Yang KF, Lu XZ. Cryopreservation of protoplasts of the alga *Porphyra yezoensis* by vitrification. Plant Science. 2004;**166**(1):97-102. DOI: 10.1016/j.plantsci.2003.08.014

[12] Liu YG, Wang XY, Liu LX. Analysis of genetic variation in surviving apple shoots following cryopreservation by vitrification. Plant Science. 2004;**166**(3):677-685. DOI: 10.1016/j.plantsci.2003.11.003

[13] Martinez D, Revilla MA, Espina A, Garcia JR. Differential scanning calorimetry applied to the storage at ultra-low temperatures of olive and hop in vitro grown shoot-tips. Thermochimica Acta. 2000;**349**:147-151. DOI: 10.1016/S0040-6031(99)00508-0

[14] Köpnick C, Grübe M, Stock J, Senula A, Mock H-P, Nagel M. Changes of soluble sugars and ATP content during DMSO droplet freezing and PVS3 droplet vitrification of potato shoot tips. Cryobiology. 2018;**85**:79-88. DOI: 10.1016/j.cryobiol.2018.09.005

[15] Nadarajan J, Akter A. Optimization of cryopreservation protocols for zygotic embryos of *Citrus reticulata*. Acta Horticulturae. 2019:137-144. DOI: 10.17660/ActaHortic.2019.1234.18

[16] Zámečník J, Faltus M, Bilavčík A. Application of differential thermal analysis for the improved regeneration after cryopreservation of vegetatively-propagated plants. Acta Horticulturae. 2019;**1234**:57-64. DOI: 10.17660/ActaHortic.2019.1234.7

[17] Vertucci CW, Stushnoff C. The state of water in acclimating vegetative buds of apple and Amelanchier and its relationship to winter hardiness. Physiologia Plantarum. 1992;**86**:503-511. ISSN: 0031-9317

[18] Wang MR, Chen L, Teixeira da Silva JA, Volk GM, Wang Q-C. Cryobiotechnology of apple (Malus spp.): Development, progress and future prospects. Plant Cell Reports. 2018;**37**:689. DOI: 10.1007/s00299-018-2249-x

[19] Keller ERJ, Senula A. Recent aspects of Allium cryopreservation in the federal German genebank. Acta Horticulturae. 2016;**1143**:35-44. DOI: 10.17660/ActaHortic.2016.1143.6

[20] Buitink J, Walters-Vertucci C, Hoekstra FA, Leprince O. Calorimetric properties of dehydrating pollen: Analysis of a desiccation tolerant and an intolerant species. Plant Physiology. 1996;**111**:235-242. DOI: 10.1104/pp.111.1.235

[21] Keller ERJ, Senula A. Recent aspects of Allium cryopreservation in the federal German genebank. In: 7th International Symposium on Edible Alliaceae; 21-25 May 2015; Nigde, Turkey. International Society for Horticultural Science. Acta Horticulturae. 2016;**1143**:35-44. DOI: 10.17660/ActaHortic.2016.1143.6

[22] Vertucci CW. Effects of cooling rate on seeds exposed to liquid nitrogen temperatures. Plant Physiology. 1989;**90**:1478-1485. DOI: 10.1104/pp.90.4.1478

[23] Vertucci CW. Relationship between thermal transitions and freezing injury in pea and soybean seeds. Plant Physiology. 1989;**90**:1121-1128. DOI: 10.1104/pp.90.3.1121

[24] Prasongsom S, Thammasiri K, Narangajavana J, Thitamadee S, Chuenboonngarm N, Panvisavas N. Vitrification-based cryopreservation of *Dendrobium cruentum* Rchb. f. seeds. Acta Horticulturae. 2019;**1234**:157-166. DOI: 10.17660/ActaHortic.2019.1234.21

[25] Kaczmarczyk A, Rokka VM, Keller ERJ. Potato shoot tip cryopreservation. A review. Potato Research. 2011;**54**:45-79. DOI: 10.1007/s11540-010-9169-7

[26] Zamecnikova J, Fernandez E, Viehmannova I, Zamecnik J, Faltus M. Preparation of shoot tips by sucrose and dehydration pre-treatment of *Ullucus tuberosus* CAL. For cryopreservation. Acta Horticulturae. 2011;**908**:331-338. DOI: 10.17660/ActaHortic.2011.908.44

[27] Gonzalez-Benito ME, Mendoza-Condori VH, Molina-Garcia AD. Cryopreservation of in vitro shoot apices of *Oxalis tuberosa* Mol. Cryo Letters. 2007;**28**:23-32. ISSN: 0143-2044

[28] Zamecnik J, Faltus M, Kotkova R, Hejnak V. Glass transition determination in allium shoot tips after dehydration. Acta Horticulturae. 2011;**908**:33-38. DOI: 10.17660/ActaHortic.2011.908.1

[29] Lynch PT, Graham RS, Zámečník J, Harding K. Optimization of water content for the cryopreservation of *Allium sativum* in vitro cultures by encapsulation-dehydration. Cryo

Letters. 2016;**37**:308-317. ISSN: 0143-2044

[30] Vogiatzi C, Grout BW, Wetten A, Toldam-Andersen BT. Cryopreservation of winter-dormant apple buds: II - tissue water status after desiccation at −4C and before further cooling. Cryo Letters. 2011;**32**:367-376. ISSN: 0143-2044

[31] Bachiri Y, Song GQ, Plessis P, Shoar-Ghaffari A, Rekab T, Morisset C. Routine cryopreservation of kiwifruit (Actinidia spp) germplasm by encapsulation-dehydration: Importance of plant growth regulators. Cryo Letters. 2001;**22**:61-74. ISSN: 0143-2044